Sir Thomas Stevenson

A Treatise on Alcohol

Sir Thomas Stevenson

A Treatise on Alcohol

ISBN/EAN: 9783744666817

Printed in Europe, USA, Canada, Australia, Japan

Cover: Foto ©berggeist007 / pixelio.de

More available books at **www.hansebooks.com**

A TREATISE

ON

ALCOHOL,

WITH

TABLES OF SPIRIT-GRAVITIES.

BY

THOMAS STEVENSON, M.D. Lond.,

Fellow of the Royal College of Physicians of London ;
Examiner in Forensic Medicine in the University of London ;
Lecturer on Chemistry and on Medical Jurisprudence at Guy's
Hospital ; One of the Official Analysts of the Home Office.
Fellow of the Institute of Chemistry.

SECOND EDITION.

LONDON:

GURNEY & JACKSON, 1, PATERNOSTER ROW.

(SUCCESSORS TO MR. VAN VOORST.)

1888.

ALERE FLAMMAM.

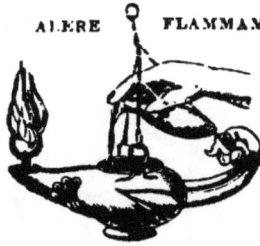

PRINTED BY TAYLOR AND FRANCIS,
RED LION COURT, FLEET STREET.

PREFACE.

In issuing a second edition of this book—originally published under the title of ' Spirit-Gravities '—I have felt it necessary to re-write the whole work, in order to bring in an account of the recent investigations, chiefly of Messrs. Squibb, on absolute alcohol. It is not necessary, however, to alter the Tables, as these are sufficiently accurate for all practical purposes; and the percentage of "proof-spirit," so necessary to the public analyst, the brewer, the distiller, and the officers of Excise and Customs, are of course unaffected by the new determinations of the specific gravity of absolute alcohol.

Chemical Laboratory,
   Guy's Hospital,
     April, 1888.

PREFACE TO THE FIRST EDITION.

THERE has long been felt a need of a convenient Table of Spirit-gravities sufficiently extended to meet the requirements of Public Analysts and others engaged in the analysis of alcoholic liquors. Existing tables are either too incomplete, or, I may add, too inaccurate, to be generally useful. This want I have endeavoured to supply. I trust that the tables which I now place before the public may be found accurate, sufficiently extended, and generally useful. They are founded upon the best accessible data, and are preceded by an historical statement relative to spirit-gravities which may, I trust, be found interesting. It should be studied by all those who may do me the honour of consulting my book.

THOMAS STEVENSON.

Chemical Laboratory,
Guy's Hospital,
August, 1880.

ALCOHOL,

INTRODUCTION TO THE TABLES.

———◆———

THE determination of the strength of an alcoholic liquid, or, as it is usually termed, an estimation of its " strength," is an operation of great importance, and one which the analyst is frequently called upon to perform. The duties levied on spirituous liquors are assessed according to the percentage of alcohol which they are supposed to contain, and hence the Governments of different countries have from time to time sought the aid of chemists and physicists in determining the strengths, expressed in terms of some given standard of alcohol. Very laborious researches have consequently been undertaken; and yet, even up to this time, the problem is so complex that it has not been altogether successfully solved. Until the law is ascertained, according to which mixtures of alcohol and water contract when mixed, we can only rely upon actual experiments, checked by well-known mathematical methods, for the construction of what are known as " spirit-gravities."

At the outset we are met by the initial diffi-

culty, that pure dry alcohol (absolute alcohol ;
ethyl-alcohol) is a substance procurable only with
extreme care and labour ; and, indeed, it was
only after the first edition of this book was published that really pure dry alcohol was obtained
by Messrs. Squibb. Even when procured, alcohol
is so highly hygroscopic that it rapidly becomes
diluted with water when exposed to any atmo-
sphere containing aqueous vapour.

The best experimental results as to the specific
gravity of pure ethyl-alcohol are as follows :—

Fownes found :

Sp. gr. ·7938 at 60° Fahr. ;

and it is probable, though not certain, that he
states the weight of alcohol at 60° compared
with the weight of an equal bulk of water, also
at 60° F.

Mendelejeff found :

Sp. gr. ·79367 at 15° C. (59° F.),

water being taken as unity at its maximum
density (4° C.) ; and as he also determined the
rate of expansion of alcohol at the given temperatures, his figures, coupled with the best determinations of the rates of expansion of water at
various temperatures, enable us to calculate the
sp. gr. of alcohol at 60° F. to be ·79404 compared with water, taken as unity, also at 60° F.

Dupré and Page found :

Sp. gr. ·79317 at 15°·5 C. (59°·9 F.),

water at 4° C. being taken as unity ; and as
they also ascertained the rate of expansion of
alcohol at various temperatures, we can calculate

the sp. gr. at 60° F. (water at 60° F. = 1) as 79391.

Drinkwater found :

Sp. gr. ·79381 at 60° F. (water at 60° F. = 1).

Messrs. Squibb (' Ephemeris,' 1884 ; ' Chemical News,' 1885, vol. lxi. pp. 5, 21, 33) have obtained alcohol of a lower specific gravity than any previous observer had recorded. Their product had its specific gravity ·79350 at 60° F., compared with water taken as unity at the same temperature, no other correction for the buoyancy of the air being made than by the use of a counterpoising flask.

According to their data, an alcohol of ·7938 true sp. gr. at 60° F. would contain 99·9 per cent. of alcohol by weight ; and proof spirit will contain 49·14, and not 49·24, per cent. of alcohol, as calculated by Drinkwater.

Absolute Alcohol. (Squibb.)

True sp. gr. at 60° F., ·79350 ; water at 60° F. = 1.

True sp. gr. (relative density) at 4° C., ·80257 ; water at 4° C.=1.

Absolute density, ·80258 (1 cub. cent. water at 4° C.=1·000013 grammes).

Apparent sp. gr. at 60° F., ·79301 ; water at 4° C.=1.

Sp. gr. at 60° F., water at 4° C. = 1, corrected for expansion of glass, ·79279.

Coefficient of expansion, when compared with water at 4°, ·0008266 for each degree Cent. at 60° F., or ·0004592 for each degree Fahr.

Percentages of real alcohol.	True sp. gr. at 60° F.	Percentages of real alcohol.	True sp. gr. at 60° F.
100	·79350	72	·86711
99	·79669	68	·87665
98	·79967	64	·88636
96	·80558	60	·89561
92	·81684	56	·90465
88	·82755	52	·91365
84	·83775	48	·92247
80	·84779	44	·93101
76	·85749	40	·93923

—Squibb.

It may therefore be stated that alcohol has approximately the following true specific gravities, adopting Squibb's figure for dry alcohol :—

Temperatures.	Water at 4° C.=1.	Water at temp. of sp. gr.=1.	Water at 60° F.=1.
15½° C.	·7930	·7935	·7935
60° F..........	·7928	·7935	·7935
62° F..........	·7919	·7928	·7926

I have not taken the decimals beyond four places, as the fifth places are unreliable.

The percentage of alcohol in a distilled liquid is best determined by observing its specific gravity at a definite temperature. In modern scientific memoirs it is customary to state the specific gravity of a liquid, with reference to water taken as

unity at its point of maximum density (4° C.).
In technical work, however, it is better, and
customary, to give the apparent specific gravity of
a liquid in terms expressing the weight of the
liquid, as compared with an equal volume of
water taken as unity at the same temperature of
60° F., which is nearly equal to 15°·6 C., no
allowance being made for the buoyancy of the air.
Spirits are, nevertheless, gauged by the hydrometer
at 62° F. (16⅔° C.), at which temperature the
imperial gallon of water weighs 10 lbs.

The first, and perhaps the most important,
scientific inquiry into the best method of determining the strength of spirituous liquors was one
undertaken by Blagden and Gilpin on the application of the English Government, and the results
were published in the ' Philosophical Transactions'
for the years 1790, 1792, and 1794. Blagden
(who devised the experiments) was Secretary, and
Gilpin (who carried them out) was Clerk, to the
Royal Society of London. The accuracy and
fidelity with which the work was executed, and
the refinement of the methods employed, have
called forth the warmest admiration and praise
from those who have taken the pains to examine
their work, which would do honour to the more
refined science of our day. Fownes's experiments,
made more than half a century later, are not for
one moment to be compared as to accuracy with
the earlier experiments of Gilpin. It has been
objected to Gilpin's experiments that absolute
alcohol was not employed. In this, and until
dry alcohol had been obtained, there was, however, an advantage, in consequence of the greater

case with which a slightly diluted spirit is mani-
pulated. Blagden and Gilpin were fully alive to
the advisability of levying a duty upon the actual
percentage of alcohol contained in any spirit.

For an account of Blagden and Gilpin's experiments I must refer my readers to the original
memoirs of those investigators (Phil. Trans. 1790,
p. 321; 1792, pp. 425, 439; 1794, p. 275). The
tables in the 'Transactions' for 1794 extend
over 112 quarto pages, each page containing
16 columns, and each column 50 lines. The
work of verifying 270,000 figures must have
been most laborious. Yet subsequent observers
and calculators have, one and all, vouched for
the correctness of the tables.

Gilpin's experiments were made by direct
weighings, every possible source of error which
even the science of the present day points out
being avoided, except that the weighings were
not reduced to a vacuum; but this source of error
usually affects the fourth decimal in a gravity to
the extent of one unit only. The fundamental data
of Blagden and Gilpin's tables are as follows:—

Sp. gr. of alcohol used ·82514,
     ,,   ,,   ,,   reduced by calculation
in all tables to an alcohol of ·825 sp. gr.
at 60° F., water at 60° F. being taken
as unity.

In the year 1811, Tralles, the Secretary of the
Academy of Sciences of Berlin, undertook, at the
request of the Prussian Government, an investi-
gation with the view of establishing a convenient,
just, and accurate mode of determining the relative duties of spirituous liquors, the results of
which furnished the basis of the method which

has been carried into operation by Prussia, France, Austria, Belgium, and Sweden.

Tralles was led to examine the reports of Blagden and Gilpin, and he pays the highest compliment to the accuracy of their work. " They [the experiments] are," he says, " more than sufficiently exact;" and he determined not to recommend to the Prussian Government, as a basis, other experiments less complete and authentic. He remarks, as Berzelius has done, that Blagden and Gilpin had omitted to take into account the air displaced by the liquids weighed. Still, this influence is without importance in practical application, for the specific gravity of an ordinary spirit reduced to a vacuum is scarcely changed by ·0001. Since Blagden and Gilpin give specific gravities to ·00001, their fifth places of decimals are unreliable ; their fourth decimal places may be correct within one unit.

In Tralles's day no chemist had been able to produce alcohol of less density than that obtained by Lowitz ; and Tralles took this as his standard alcohol. It had the following specific gravity :—

·791 at 15° R. (68° F.), water = 1· at the same temperature.

If this sp. gr. be compared with that of water at 60° F., we get for Tralles's alcohol—

·78962 at 68° F., water at 60° F. = 1,

or ·7942 at 60°, water being unity at the same temperature. Tralles, however, appears— as Gilbert says, purposely—to have assumed a smaller rate of expansion than he himself found for his alcohol by experiment ; and, owing to this error, he takes the specific gravity of his alcohol at

60° as ·7946, compared with water taken as unity at 60°, or ·7939, compared with water at its point of maximum density (4° C.). The sp. gr. ·7946, adopted by Tralles, is often referred to as a well-determined result; but it is not so, and its accuracy appears to be more than doubtful. Tralles having determined to adopt the work of Gilpin, it became necessary to determine how much water is contained in alcohol of the specific gravity ·825 (Gilpin's alcohol). By a series of experiments this was found to be 92·6 per cent. of Tralles's alcohol by volume, or 89·2 per cent. by weight. The true percentage of absolute alcohol in Gilpin's alcohol is 88·96 per cent. by weight, or 92·49 per cent. by volume.

The table of Tralles, found in many textbooks, gives in one column the specific gravities at 60° F. (water at 4° C. = 1), and in another the percentages of alcohol by volume. The range is from pure water, sp. gr. ·9991, to alcohol, sp. gr. ·7939.

Tralles did not publish the experiments from which he deduced his conclusion that Gilpin's alcohol contained 89·2 of Tralles's alcohol; and Gilbert says that the data were not among the papers confided to him. Dividing Tralles's gravities by ·9991, to make them comparable with Gilpin's, and assuming that Tralles's alcohol contained $100 \times \dfrac{88·96}{89·2} = 99·73$ per cent. of alcohol by weight or 99·85 per cent. by volume, I compared Tralles's table with Gilpin's at the several points at which their specific gravities are comparable; and I find the annexed results at the points where the specific gravities are practically coincident.

GILPIN.		TRALLES.			
Sp. gr. at 60° F. Water=1 at 60° F.	Percentage of alcohol by volume (calculated).	Sp. gr. at 60° F. Water=1 at 4° C.	Calculated sp. gr. at 60° F. Water=1 at 60° F.	Percentage of alcohol by volume.	Excess or deficit as compared with Gilpin's experiments.
·96880	26·95	·9679	·96877	27·	+0·05
·96311	32·06	·9622	·96307	32·	—0·07
·94696	42·98	·9461	·94695	43·	+0·01
·92225	55·92	·9213	·92213	56·	+0·07
·88505	71·94	·8842	·88500	72·	+0·05
·85265	83·89	·8518	·85257	84·	+0·09
·82731	91·70	·8265	·82724	92·	+0·28

It is evident that, with percentages of alcohol by volume up to about 84 per cent., the results of Gilpin and of Tralles coincide within about one tenth of one per cent. Above 85 per cent. the results are no longer comparable. From other data I calculate that Tralles's alcohol contained 99·77 per cent. alcohol by weight, and 99·89 per cent. by volume.

Fownes's Experiments.—In the ' Philosophical Transactions ' for 1847, p. 249, is a paper by Fownes " On the Value in Absolute Alcohol of Spirits of different Specific Gravities," in which the results are arranged in a tabular form. This table is met with in most text-books of chemistry. Fownes used an absolute alcohol of sp. gr. 0·7938 at 60° F. The spirit-mixtures were made by weighing out the alcohol and water in the required proportions, and mixing in stoppered bottles. The gravities were taken after the lapse of three or four days. The following abridged Table gives all Fownes's *experimental* numbers. Intermediate percentage and gravities were obtained by calculation :—

Percentages of real alcohol.	Sp. gr. at 60° F.	Percentages of real alcohol.	Sp. gr. at 60° F.
2	·9965	52	·9135
4	·9930	54	·9090
6	·9898	56	·9047
8	·9869	58	·9001
10	·9841	60	·8956
12	·9815	62	·8908
14	·9789	64	·8863
16	·9766	66	·8816
18	·9741	68	·8769
20	·9716	70	·8721
22	·9691	72	·8672
24	·9665	74	·8625
26	·9638	76	·8581
28	·9609	78	·8533
30	·9578	80	·8483
32	·9544	82	·8434
34	·9511	84	·8382
36	·9470	86	·8331
38	·9434	88	·8279
40	·9396	90	·8228
42	·9356	92	·8172
44	·9314	94	·8118
46	·9270	96	·8061
48	·9228	98	·8001
50	·9184	100	·7938

—Fownes.

Fownes gives no information as to the accuracy
of his weighings, nor the reductions which he
made to obtain the figures of his table from the
experiments.   He does not state whether any
correction was made for air-weighings, so as to
reduce the results to weights *in vacuo* ; nor does
he state the unit of density that he takes—
whether water at 4° C. or at 60° F.   He omits
also all mention of the exactitude of his ther-
mometry, or of the means employed to weigh the
spirit at the exact temperature employed.   Hence
the reputation of Fownes has to be accepted in
lieu of a detailed narrative of actual experimental
facts.   We have no assurance of the accuracy of
Fownes's table, and it appears to me to be far
inferior in accuracy to the tables of Blagden and
Gilpin.   Prof. R. S. McCulloh, in an exhaustive
report to the Secretary of the Treasury, U.S.,
in 1848 *, speaks, too, in disparaging terms of
Fownes's work.   Yet Mr. Hehner, who has ela-
borated a set of tables of spirit-gravities, based on
Fownes's tables, terms these " the excellent tables
of Fownes; " and adds, " all later investigators
have confirmed the general accuracy of Fownes's
table."   I have been unable to ascertain by
whom the table referred to has been verified in
detail; and I am assured, on competent authority,
that Mr. Hehner's tables are incorrect at some

* Reports from the Secretary of the Treasury of Scientific Investigations in relation to Sugar and Hydro-
meters, made under the superintendence of Prof. A. D.
Bache by Prof. R. S. McCulloh, Washington, 1848.   Re-
vised by order of the Senate, 30th Congress, 1st Session,
Ex. Doc. No. 50.

points to the extent of nearly one half per cent.
of proof spirit—a serious difference. Fownes's
results differ greatly from those of Blagden and
Gilpin, on which are based all the tables of
the British Excise, the two being discrepant at
certain points to the extent of more than one
third of one per cent. of proof spirit. And,
moreover, it must be borne in mind that suc-
cessive experimenters have one and all vouched
to the accuracy of Gilpin's tables.

Proof Spirit.

Proof spirit is defined by Statute (58 Geo. III.
c. 28) as "that which, at a temperature of fifty-
one degrees by Fahrenheit's thermometer, weighs
exactly twelve thirteenths of an equal measure
of distilled water." It is assumed, though not
enacted, that the water is likewise at the temperature of 51° F. Hence the specific gravity of
proof spirit at 51° F. is ·92308, when compared
with water at the same temperature. An *experimental* determination of its specific gravity at
60° F. or any other temperature than 51° F. has
never, so far as I know, been made. Drinkwater, in the year 1847, communicated to the
Chemical Society of London (Phil. Mag. vol. xxii.
1848, p. 123) his calculation of the specific
gravity of proof spirit at 60° F., based upon
Gilpin's experiments; and this he estimated to
be ·91984 compared with water taken as unity
at the same temperature. Since Gilpin's experimental determinations of the density of water at

51° and 60° have been confirmed by the best modern observers, the sp. gr. ·91984 may be accepted as correct. Operating upon diluted spirt of wine containing 49 and 49·5 respectively of alcohol, sp. gr. ·7938 at 60° F., Drinkwater got *by interpolation* 49·24 per cent. by weight as the percentage of alcohol in proof spirit ; and this corresponds to 57·06 per cent. by volume. These numbers, corrected for absolute alcohol of sp. gr. ·7935 at 60° F., become 49·19 and 57·01 respectively.

Drinkwater also determined the specific gravity of mixtures of alcohol and water, both at 60° F. In determining these specific gravities he took minute precautions to avoid errors. I have compared Drinkwater's tables with the earlier ones of Gilpin, and find them in remarkable agreement. They differ, however, considerably from Fownes's determinations—generally to the extent of one third of one per cent. of proof spirit. Unfortunately Drinkwater did not carry his experiments beyond mixtures containing, as a maximum, ten per cent. of proof spirit.

My own following tables are calculated from Gilpin's tables in the ' Philosophical Transactions of the Royal Society of London ' for 1794. Gilpin's tables extend to spirits ranging in specific gravity from ·8250 to ·9983, or from spirit containing 88 by weight of alcohol to ·9 per cent. These have been compared, between the specific gravities ·9839 and ·9983, with those of Drinkwater. Gilpin's and Drinkwater's experiments are in complete accordance. For spirits of less specific gravity than ·8250 the data of Fownes, in the absence of better experiments, were adopted. The want of reliability in Fownes's data is of less importance, since spirits of greater alcoholic strength than 89 per cent. (62 over proof) are rarely met with outside the laboratory. The specific gravities proceed by equal increments of ·0001, *i. e.* of one unit in the fourth decimal place. Opposite the specific gravity in the second column is found the percentage, weight in weight, in alcohol of the stated specific gravity, no reduction being made to reduce the air-weighings to the true weighings *in vacuo.* This reduction does not much affect, however, the fourth decimal place, and hence may be neglected. In the third column is the percentage by volume of absolute alcohol. In the fourth column is the corresponding percentage of proof spirit. The specific gravities are those of the spirit at 60° F. ($15\frac{5}{9}$° C.), compared with water at the same temperature.

In issuing a second edition of this work it has not been thought necessary to reduce the calculations made upon alcohol of ·7938 specific gravity to an alcohol of ·7935 specific gravity— the lowest gravity hitherto obtained for absolute alcohol. The difference is inconsiderable. If $\frac{1}{1000}$ be deducted from the figures in the second column, the percentage of alcohol sp. gr. ·7938 will be obtained. Thus spirit of ·8228 sp. gr. has in the second column 90· as its percentage of alcohol (·7938 sp. gr.), and $90· - ·09 = 89·91$ is the percentage of alcohol of sp. gr. ·7935. The correction is easily made mentally if required.

In the second and third columns the per-

centages are given to the nearest $\frac{1}{20}$ (0·05) per
cent. ; and in the fourth column to the nearest
$\frac{1}{10}$ (·1) per cent. Thus, when a spirit was
found to have a percentage of alcohol by weight
56·12, it is given as 52·10 per cent. ; and a spirit
of 56·13 per cent. is stated as 56·15 per cent.
A proof spirit, strength 25·03 per cent., is given
as 25·0 per cent. ; and one of 25·06 as 25·1 per
cent. $\frac{1}{20}$ per cent. of alcohol and $\frac{1}{10}$ per
cent. of proof spirit are the limits of ordinary
accurate working. Tables giving second decimal
places differing by units have only an illusory
appearance of minute accuracy.

CORRECTIONS FOR TEMPERATURE.

The expansion of a mixture of alcohol and
water with rise of temperature is very regular
for all such mixtures ; but each diluted spirit
has its specific rate of expansion. Should the
specific gravity of a spirit be taken at any other
temperature than 60° F., a correction must be
made in order to obtain the true specific gravity
at 60° F. The following are sufficiently near
approximations, for all ordinary purposes, for
temperatures between 55° and 68° F. ($12\frac{3}{4}$° and
20° C.) :—

Alcohol 5-15°/₀ by vol. add or deduct ·0001 to or from sp. gr.
,, 15-25°/₀ ,, ,, ,, ·0002 ,, ,,
,, 25-40°/₀ ,, ,, ,, ·0003 ,, ,,
,, 40-70°/₀ ,, ,, ,, ·0004 ,, ,,
,, above 70°/₀ ,, ,, ,, ·0005 ·, ,,

for each 1° F. out.

These quantities must be added when the temperature at which the specific gravity was taken is above 60° F., and deducted when the temperature of observation was below 60° F.

A corresponding correction may be applied for each 0°·5 C. excess or deficit of temperature if a Centigrade thermometer be used.

USEFUL DATA.

Absolute Alcohol :—

> Sp. gr. ·7935 at 60° F. ; water at 60° F.=1.
> „ „ ·7928 „ 60° „ „ „ 4° C.=1.
> „ „ ·7926 „ 62° „ „ „ 60° F.=1.
> „ „ ·7928 „ 62° „ „ „ 62° F.=1.
> „ „ ·7919 „ 62° „ „ „ 4° C.=1.

Proof Spirit :—

> 49·19 per cent. alcohol by weight +50·81 per cent. water.
> 57·01 „ „ „ volume.

> Sp. gr. ·9231 at 51° F. ; water at 51° F.=1.
> „ „ ·9198 „ 60° „ „ „ 60° F.=1.
> „ „ ·9189 „ 60° „ „ „ 4° C.=1.
> „ „ ·9190 „ 62° „ „ „ 60° F.=1.
> „ „ ·9180 „ 62° „ „ „ 4° C.=1.
> „ „ ·9192 „ 62° „ „ „ 62° F.=1.

Blagden and Gilpin's Spirit :—

> 88·96 per cent. of alcohol by weight +11·04 per cent. water.
> 92·49 „ „ „ volume.

> Sp. gr. ·8250 at 60° F. ; water at 60° F.=1.
> „ „ ·8242 „ 60° „ „ „ 4° C.=1.
> „ „ ·8240 „ 62° „ „ „ 60° F.=1.
> „ „ ·8232 „ 62° „ „ „ 4° C.=1.

> Per cent. by weight ÷1·124=per cent. real alc. by wt.
> „ „ volume÷1·081= „ „ „ vol.

Tralles's Alcohol :—

> 99·7 per cent. alcohol by weight +0·3 per cent. water.
> 99·85 „ „ „ volume.

> Sp. gr. ·7946 at 60° F. ; water at 60° F.=1.
> „ „ ·7939 „ 60° „ „ „ 4° C.=1.

> Per cent. by weight ÷1·003=per cent. real alc. by wt.
> „ „ volume÷1·002= „ „ „ vol.

I. Sp. gr. at 0° F. Water at 60°=1.	II. Percentage of alcohol *by weight.*	III. Percentage of alcohol *by volume.*	IV. Percentage of Proof Spirit.
·7938	100·00	100·00	175·25
·7939	99·95	99·95	175·2
·7940	99·95	99·95	175·2
·7941	99·90	99·95	175·2
·7942	99·90	99·90	175·1
·7943	99·85	99·90	175·1
·7944	99·80	99·90	175·0
·7945	99·80	99·85	175·0
·7946	99·75	99·85	174·9
·7947	99·70	99·80	174·9
·7948	99·70	99·80	174·9
·7949	99·65	99·80	174·8
·7950	99·60	99·75	174·8
·7951	99·60	99·75	174·7
·7952	99·55	99·70	174·7
·7953	99·55	99·70	174·7
·7954	99·50	99·70	174·6
·7955	99·45	99·65	174·6
·7956	99·45	99·65	174·6
·7957	99·40	99·60	174·5
·7958	99·35	99·60	174·5
·7959	99·35	99·60	174·4
·7960	99·30	99·55	174·4
·7961	99·30	99·55	174·4
·7962	99·25	99·50	174·3
·7963	99·20	99·50	174·3
·7964	99·20	99·50	174·2
·7965	99·15	99·45	174·2

I. Sp. gr. at 60° F. Water at 60°=1.	II. Percentage of alcohol by weight.	III. Percentage of alcohol by volume.	IV. Percentage of Proof Spirit.
·7966	99·10	99·45	174·2
·7967	99·05	99·45	174·1
·7968	99·00	99·40	174·1
·7969	99·00	99·40	174·1
·7970	98·95	99·40	174·1
·7971	98·95	99·35	174·0
·7972	98·90	99·35	174·0
·7973	98·90	99·30	174·0
·7974	98·85	99·30	173·9
·7975	98·85	99·30	173·9
·7976	98·80	99·25	173·9
·7977	98·75	99·25	173·8
·7978	98·75	99·20	173·8
·7979	98·70	99·20	173·8
·7980	98·65	99·20	173·8
·7981	98·65	99·15	173·7
·7982	98·60	99·15	173·7
·7983	98·55	99·10	173·7
·7984	98·55	99·10	173·6
·7985	98·50	99·10	173·6
·7986	98·50	99·05	173·6
·7987	98·45	99·05	173·5
·7988	98·40	99·00	173·5
·7989	98·40	99·00	173·5
·7990	98·35	99·00	173·5
·7991	98·30	98·95	173·4
·7992	98·30	98·95	173·4
·7993	98·25	98·90	173·4

I.	II.	III.	IV.
Sp. gr. at 60° F. Water at 60°=1.	Percentage of alcohol *by weight.*	Percentage of alcohol *by volume.*	Percentage of Proof Spirit.
·7994	98·20	98·90	173·3
·7995	98·20	98·85	173·3
·7996	98·15	98·85	173·3
·7997	98·15	98·85	173·2
·7998	98·10	98·80	173·2
·7999	98·05	98·80	173·2
·8000	98·00	98·75	173·1
·8001	98·00	98·75	173·1
·8002	97·95	98·75	173·1
·8003	97·95	98·70	173·0
·8004	97·90	98·70	173·0
·8005	97·85	98·65	172·9
·8006	97·85	98·65	172·9
·8007	97·80	98·65	172·8
·8008	97·75	98·60	172·8
·8009	97·75	98·60	172·7
·8010	97·70	98·55	172·7
·8011	97·65	98·55	172·7
·8012	97·65	98·55	172·6
·8013	97·60	98·50	172·6
·8014	97·55	98·50	172·6
·8015	97·55	98·45	172·5
·8016	97·50	98·45	172·5
·8017	97·45	98·45	172·5
·8018	97·45	98·40	172·4
·8019	97·40	98·40	172·4
·8020	97·35	98·40	172·4
·8021	97·35	98·35	172·3

I. Sp. gr. at 60° F. Water at 60°=1.	II. Percentage of alcohol *by weight.*	III. Percentage of alcohol *by volume.*	IV. Percentage of Proof Spirit.
·8022	97·30	98·35	172·3
·8023	97·25	98·30	172·3
·8024	97·25	98·30	172·2
·8025	97·20	98·25	172·2
·8026	97·15	98·25	172·2
·8027	97·15	98·25	172·1
·8028	97·10	98·20	172·1
·8029	97·05	98·20	172·1
·8030	97·05	98·15	172·0
·8031	97·00	98·15	172·0
·8032	96·95	98·10	172·0
·8033	96·95	98·10	171·9
·8034	96·90	98·05	171·9
·8035	96·85	98·05	171·8
·8036	96·85	98·00	171·8
·8037	96·80	98·00	171·8
·8038	96·75	98·00	171·7
·8039	96·75	97·95	171·7
·8040	96·70	97·95	171·6
·8041	96·65	97·90	171·6
·8042	96·65	97·90	171·6
·8043	96·60	97·90	171·5
·8044	96·55	97·85	171·5
·8045	96·55	97·85	171·4
·8046	96·50	97·80	171·4
·8047	96·45	97·80	171·4
·8048	96·45	97·75	171·3
·8049	96·40	97·75	171·3
·8050	96·35	97·70	171·2

4

I. Sp. gr. at 60° F. Water at 60°=1.	II. Percentage of alcohol *by weight.*	III. Percentage of alcohol *by volume.*	IV. Percentage of Proof Spirit.
·8051	96·35	97·70	171·2
·8052	96·30	97·70	171·2
·8053	96·25	97·65	171·1
·8054	96·25	97·65	171·1
·8055	96·20	97·60	171·0
·8056	96·15	97·60	171·0
·8057	96·15	97·60	171·0
·8058	96·10	97·55	170·9
·8059	96·05	97·55	170·9
·8060	96·05	97·50	170·8
·8061	96·00	97·50	170·8
·8062	95·95	97·45	170·8
·8063	95·95	97·45	170·7
·8064	95·90	97·40	170·7
·8065	95·85	97·40	170·6
·8066	95·80	97·35	170·6
·8067	95·80	97·35	170·5
·8068	95·75	97·30	170·5
·8069	95·70	97·30	170·5
·8070	95·70	97·25	170·4
·8071	95·65	97·25	170·4
·8072	95·60	97·20	170·3
·8073	95·60	97·20	170·3
·8074	95·55	97·15	170·3
·8075	95·50	97·15	170·2
·8076	95·45	97·15	170·2
·8077	95·45	97·10	170·1
·8078	95·40	97·10	170·1
·8079	95·35	97·05	170·0

I. Sp. gr. at 60° F. Water at 60°=1.	II. Percentage of alcohol *by weight.*	III. Percentage of alcohol *by volume.*	IV. Percentage of Proof Spirit.
·8080	95·30	97·05	170·0
·8081	95·30	97·00	170·0
·8082	95·25	97·00	169·9
·8083	95·20	96·95	169·9
·8084	95·20	96·95	169·8
·8085	95·15	96·90	169·8
·8086	95·10	96·90	169·7
·8087	95·05	96·85	169·7
·8088	95·05	96·85	169·6
·8089	95·00	96·80	169·6
·8090	94·95	96·80	169·6
·8091	94·95	96·75	169·5
·8092	94·90	96·75	169·5
·8093	94·85	96·70	169·5
·8094	94·85	96·70	169·4
·8095	94·80	96·65	169·4
·8096	94·75	96·65	169·3
·8097	94·75	96·60	169·3
·8098	94·70	96·60	169·2
·8099	94·65	96·60	169·2
·8100	94·60	96·55	169·2
·8101	94·60	96·55	169·1
·8102	94·55	96·50	169·1
·8103	94·55	96·50	169·1
·8104	94·50	96·45	169·0
·8105	94·45	96·45	169·0
·8106	94·45	96·40	169·0
·8107	94·40	96·40	168·9

I. Sp. gr. at 60° F. Water at 60°=1.	II. Percentage of alcohol *by* *weight*.	III. Percentage of alcohol *by* *volume*.	IV. Percentage of Proof Spirit.
·8108	94·35	96·35	168·9
·8109	94·35	96·35	168·9
·8110	94·30	96·35	168·8
·8111	94·25	96·30	168·8
·8112	94·25	96·30	168·7
·8113	94·20	96·25	168·7
·8114	94·15	96·25	168·7
·8115	94·10	96·20	168·6
·8116	94·05	96·20	168·6
·8117	94·05	96·15	168·5
·8118	94·00	96·15	168·5
·8119	93·95	96·10	168·4
·8120	93·95	96·10	168·4
·8121	93·90	96·05	168·3
·8122	93·85	96·05	168·3
·8123	93·80	96·00	168·2
·8124	93·80	96·00	168·2
·8125	93·75	95·95	168·1
·8126	93·70	95·90	168·1
·8127	93·65	95·90	168·0
·8128	93·65	95·85	168·0
·8129	93·60	95·85	167·9
·8130	93·55	95·80	167·9
·8131	93·50	95·75	167·8
·8132	93·50	95·75	167·8
·8133	93·45	95·70	167·8
·8134	93·40	95·70	167·7
·8135	93·35	95·65	167·7

I. Sp. gr. at 60° F. Water at 60°=1.	II. Percentage of alcohol *by weight*.	III. Percentage of alcohol *by volume*.	IV. Percentage of Proof Spirit.
·8136	93·35	95·65	167·6
·8137	93·30	95·60	167·6
·8138	93·25	95·60	167·5
·8139	93·20	95·55	167·5
·8140	93·20	95·50	167·4
·8141	93·15	95·50	167·4
·8142	93·10	95·45	167·4
·8143	93·05	95·45	167·3
·8144	93·05	95·40	167·3
·8145	93·00	95·40	167·2
·8146	92·95	95·35	167·2
·8147	92·95	95·35	167·1
·8148	92·90	95·30	167·1
·8149	92·85	95·30	167·0
·8150	92·80	95·25	167·0
·8151	92·80	95·25	166·9
·8152	92·75	95·20	166·9
·8153	92·70	95·20	166·8
·8154	92·65	95·15	166·8
·8155	92·65	95·15	166·8
·8156	92·60	95·10	166·7
·8157	92·55	95·10	166·7
·8158	92·50	95·05	166·6
·8159	92·50	95·05	166·6
·8160	92·45	95·00	166·5
·8161	92·40	95·00	166·5
·8162	92·35	94·95	166·4
·8163	92·35	94·95	166·4

I. Sp. gr. at 60° F. Water at 60°=1.	II. Percentage of alcohol *by weight*.	III. Percentage of alcohol *by volume*.	IV. Percentage of Proof Spirit.
·8164	92·30	94·90	166·3
·8165	92·25	94·90	166·3
·8166	92·20	94·85	166·2
·8167	92·20	94·80	166·2
·8168	92·15	94·80	166·2
·8169	92·10	94·75	166·1
·8170	92·05	94·75	166·1
·8171	92·05	94·70	166·0
·8172	92·00	94·70	166·0
·8173	91·95	94·65	165·9
·8174	91·95	94·65	165·9
·8175	91·90	94·60	165·8
·8176	91·85	94·60	165·8
·8177	91·80	94·55	165·7
·8178	91·80	94·55	165·7
·8179	91·75	94·50	165·6
·8180	91·70	94·50	165·6
·8181	91·65	94·45	165·5
·8182	91·65	94·45	165·5
·8183	91·60	94·40	165·4
·8184	91·55	94·40	165·4
·8185	91·50	94·35	165·3
·8186	91·50	94·35	165·3
·8187	91·45	94·30	165·2
·8188	91·40	94·30	165·2
·8189	91·35	94·25	165·1
·8190	91·35	94·25	165·1
·8191	91·30	94·20	165·0

I. Sp. gr. at 60° F. Water at 60°=1.	II. Percentage of alcohol *by weight*.	III. Percentage of alcohol *by volume*.	IV. Percentage of Proof Spirit.
·8192	91·25	94·20	165·0
·8193	91·20	94·15	164·9
·8194	91·20	94·15	164·9
·8195	91·15	94·10	164·9
·8196	91·10	94·10	164·8
·8197	91·05	94·05	164·8
·8198	91·05	94·05	164·8
·8199	91·00	94·00	164·7
·8200	90·95	94·00	164·7
·8201	90·95	93·95	164·7
·8202	90·90	93·95	164·6
·8203	90·85	93·90	164·6
·8204	90·85	93·90	164·5
·8205	90·80	93·85	164·5
·8206	90·75	93·85	164·4
·8207	90·75	93·80	164·4
·8208	90·70	93·80	164·4
·8209	90·65	93·75	164·3
·8210	90·65	93·75	164·3
·8211	90·60	93·70	164·3
·8212	90·55	93·70	164·2
·8213	90·50	93·65	164·2
·8214	90·45	93·65	164·1
·8215	90·45	93·60	164·1
·8216	90·40	93·60	164·1
·8217	90·35	93·55	164·0
·8218	90·35	93·55	164·0
·8219	90·30	93·50	163·9
·8220	90·25	93·50	163·9

I. Sp. gr. at 60° F. Water at 60°=1.	II. Percentage of alcohol *by weight*.	III. Percentage of alcohol *by volume*.	IV. Percentage of Proof Spirit.
·8221	90·25	93·45	163·8
·8222	90·20	93·45	163·8
·8223	90·15	93·40	163·7
·8224	90·15	93·40	163·7
·8225	90·10	93·35	163·6
·8226	90·05	93·35	163·6
·8227	90·05	93·30	163·5
·8228	90·00	93·30	163·5
·8229	89·95	93·25	163·4
·8230	89·90	93·25	163·4
·8231	89·85	93·20	163·3
·8232	89·80	93·15	163·3
·8233	89·75	93·15	163·2
·8234	89·75	93·10	163·1
·8335	89·70	93·05	163·1
·8236	89·65	93·05	163·0
·8237	89·60	93·00	163·0
·8238	89·55	92·95	162·9
·8239	89·55	92·95	162·9
·8240	89·50	92·90	162·8
·8241	89·45	92·90	162·7
·8242	89·40	92·85	162·7
·8243	89·35	92·80	162·6
·8244	89·30	92·75	162·6
·8245	89·30	92·70	162·5
·8246	89·25	92·70	162·4
·8247	89·20	92·65	162·4
·8248	89·15	92·60	162·3
·8249	89·10	92·60	162·3

I.	II.	III.	IV.
Sp. gr. at 60° F. Water at 60°=1.	Percentage of alcohol *by weight*.	Percentage of alcohol *by volume*.	Percentage of Proof Spirit.
·8250	89·05	92·55	162·2
·8251	89·00	92·55	162·2
·8252	89·00	92·50	162·1
·8253	88·95	92·50	162·1
·8254	88·90	92·45	162·0
·8255	88·85	92·45	162·0
·8256	88·80	92·40	161·9
·8257	88·75	92·40	161·9
·8258	88·70	92·35	161·8
·8259	88·70	92·35	161·8
·8260	88·65	92·30	161·7
·8261	88·60	92·30	161·7
·8262	88·55	92·25	161·6
·8263	88·55	92·20	161·6
·8264	88·50	92·15	161·5
·8265	88·45	92·15	161·5
·8266	88·45	92·10	161·4
·8267	88·40	92·10	161·4
·8268	88·40	92·05	161·3
·8269	88·35	92·00	161·2
·8270	88·30	92·00	161·2
·8271	88·25	91·95	161·1
·8272	88·20	91·95	161·1
·8273	88·15	91·90	161·0
·8274	88·15	91·90	161·0
·8275	88·10	91·85	160·9
·8276	88·05	91·80	160·9
·8277	88·00	91·80	160·8

I. Sp. gr. at 60° F. Water at 60°=1.	II. Percentage of alcohol *by* *weight.*	III. Percentage of alcohol *by* *volume.*	IV. Percentage of Proof Spirit.
·8278	88·00	91·75	160·8
·8279	87·95	91·70	160·7
·8280	87·90	91·70	160·7
·8281	87·85	91·65	160·6
·8282	87·85	91·65	160·6
·8283	87·80	91·60	160·5
·8284	87·75	91·60	160·5
·8285	87·70	91·55	160·4
·8286	87·65	91·50	160·4
·8287	87·60	91·50	160·3
·8288	87·60	91·45	160·3
·8289	87·55	91·45	160·2
·8290	87·50	91·40	160·2
·8291	87·50	91·40	160·1
·8292	87·45	91·35	160·1
·8293	87·40	91·35	160·0
·8294	87·35	91·30	160·0
·8295	87·35	91·30	159·9
·8296	87·30	91·25	159·9
·8297	87·30	91·25	159·8
·8298	87·25	91·20	159·8
·8299	87·20	91·15	159·7
·8300	87·15	91·10	159·7
·8301	87·10	91·10	159·6
·8302	87·05	91·05	159·6
·8303	87·00	91·05	159·5
·8304	87·00	91·00	159·5
·8305	86·95	91·00	159·4

I. Sp. gr. at 60° F. Water at 60°=1.	II. Percentage of alcohol *by weight*.	III. Percentage of alcohol *by volume*.	IV. Percentage of Proof Spirit.
·8306	86·90	90·95	159·3
·8307	86·85	90·95	159·3
·8308	86·80	90·90	159·2
·8309	86·80	90·90	159·2
·8310	86·75	90·85	159·1
·8311	86·70	90·80	159·0
·8312	86·70	90·80	159·0
·8313	86·65	90·75	158·9
·8314	86·60	90·75	158·9
·8315	86·55	90·70	158·8
·8316	86·50	90·70	158·8
·8317	86·45	90·65	158·7
·8318	86·45	90·60	158·7
·8319	86·40	90·60	158·6
·8320	86·35	90·55	158·6
·8321	86·30	90·50	158·5
·8322	86·30	90·50	158·5
·8323	86·25	90·45	158·5
·8324	86·20	90·40	158·4
·8325	86·20	90·40	158·4
·8326	86·15	90·35	158·3
·8327	86·10	90·30	158·3
·8328	86·05	90·30	158·2
·8329	86·00	90·25	158·2
·8330	85·95	90·25	158·1
·8331	85·90	90·20	158·1
·8332	85·85	90·20	158·0
·8333	85·80	90·15	158·0

I. Sp. gr. at 60° F. Water at 60°=1.	II. Percentage of alcohol *by weight*.	III. Percentage of alcohol *by volume*.	IV. Percentage of Proof Spirit.
·8334	85·75	90·10	157·9
·8335	85·75	90·10	157·9
·8336	85·70	90·05	157·8
·8337	85·65	90·00	157·8
·8338	85·65	90·00	157·7
·8339	85·60	89·95	157·7
·8340	85·60	89·95	157·6
·8341	85·55	89·90	157·6
·8342	85·50	89·85	157·5
·8343	85·50	89·85	157·5
·8344	85·45	89·80	157·4
·8345	85·40	89·75	157·4
·8346	85·35	89·75	157·3
·8347	85·30	89·70	157·3
·8348	85·30	89·65	157·2
·8349	85·25	89·65	157·2
·8350	85·20	89·60	157·1
·8351	85·15	89·60	157·1
·8352	85·15	89·55	157·0
·8353	85·10	89·50	156·9
·8354	85·05	89·50	156·9
·8355	85·00	89·45	156·8
·8356	85·00	89·40	156·8
·8357	84·95	89·40	156·7
·8358	84·90	89·35	156·6
·8359	84·85	89·35	156·6
·8360	84·80	89·30	156·5
·8361	84·80	89·30	156·5

I. Sp. gr. at 60° F. Water at 60°=1.	II. Percentage of alcohol *by weight.*	III. Percentage of alcohol *by volume.*	IV. Percentage of Proof Spirit.
·8362	84·75	89·25	156·4
·8363	84·75	89·20	156·4
·8364	84·70	89·15	156·3
·8365	84·65	89·15	156·3
·8366	84·60	89·10	156·2
·8367	84·55	89·10	156·2
·8368	84·50	89·05	156·1
·8369	84·45	89·00	156·1
·8370	84·40	89·00	156·0
·8371	84·35	88·95	156·0
·8372	84·30	88·90	155·9
·8373	84·25	88·90	155·8
·8374	84·20	88·85	155·8
·8375	84·20	88·85	155·7
·8376	84·15	88·80	155·6
·8377	84·10	88·75	155·6
·8378	84·10	88·75	155·5
·8379	84·05	88·70	155·4
·8380	84·00	88·65	155·4
·8381	84·00	88·60	155·3
·8382	83·95	88·60	155·3
·8383	83·90	88·55	155·2
·8384	83·85	88·55	155·2
·8385	83·80	88·50	155·1
·8386	83·75	88·45	155·1
·8387	83·70	88·45	155·0
·8388	83·70	88·40	155·0
·8389	83·65	88·40	154·9
·8390	83·60	88·35	154·9

I. Sp. gr. at 60° F. Water at 60°=1.	II. Percentage of alcohol *by weight*.	III. Percentage of alcohol *by volume*.	IV. Percentage of Proof Spirit.
·8391	83·55	88·30	154·8
·8392	83·50	88·25	154·8
·8393	83·50	88·25	154·7
·8394	83·45	88·20	154·6
·8395	83·40	88·20	154·6
·8396	83·35	88·15	154·5
·8397	83·30	88·15	154·5
·8398	83·30	88·10	154·4
·8399	83·25	88·10	154·4
·8400	83·20	88·05	154·3
·8401	83·15	88·05	154·2
·8402	83·10	88·00	154·2
·8403	83·10	87·95	154·1
·8404	83·05	87·95	154·1
·8405	83·00	87·90	154·0
·8406	83·00	87·90	154·0
·8407	82·95	87·85	153·9
·8408	82·90	87·80	153·9
·8409	82·85	87·80	153·8
·8410	82·80	87·75	153·8
·8411	82·80	87·70	153·7
·8412	82·75	87·70	153·7
·8413	82·70	87·65	153·6
·8414	82·65	87·60	153·5
·8415	82·60	87·60	153·5
·8416	82·60	87·55	153·4
·8417	82·55	87·50	153·3
·8418	82·50	87·50	153·3
·8419	82·45	87·45	153·2

c

I. Sp. gr. at 60° F. Water at 60°=1.	II. Percentage of alcohol *by weight.*	III. Percentage of alcohol *by volume.*	IV. Percentage of Proof Spirit.
·8420	82·45	87·45	153·2
·8421	82·40	87·40	153·1
·8422	82·35	87·40	153·1
·8423	82·30	87·35	153·0
·8424	82·25	87·30	153·0
·8425	82·20	87·30	152·9
·8426	82·15	87·25	152·9
·8427	82·10	87·20	152·8
·8428	82·05	87·20	152·8
·8429	82·00	87·15	152·7
·8430	82·00	87·10	152·7
·8431	81·95	87·10	152·6
·8432	81·90	87·05	152·5
·8433	81·90	87·00	152·5
·8434	81·85	87·00	152·4
·8435	81·80	86·95	152·4
·8436	81·80	86·95	152·3
·8437	81·75	86·90	152·3
·8438	81·70	86·85	152·2
·8439	81·65	86·85	152·2
·8440	81·65	86·80	152·1
·8441	81·60	86·75	152·0
·8442	81·55	86·75	152·0
·8443	81·50	86·70	151·9
·8444	81·45	86·70	151·8
·8445	81·40	86·65	151·8
·8446	81·40	86·60	151·7
·8447	81·35	86·60	151·6

I. Sp. gr. at 60° F. Water at 60°=1.	II. Percentage of alcohol *by weight*.	III. Percentage of alcohol *by volume*.	IV. Percentage of Proof Spirit.
·8448	81·30	86·55	151·6
·8449	81·25	86·50	151·5
·8450	81·20	86·50	151·5
·8451	81·15	86·45	151·4
·8452	81·15	86·40	151·4
·8453	81·10	86·40	151·3
·8454	81·05	86·35	151·2
·8455	81·00	86·30	151·2
·8456	81·00	86·30	151·1
·8457	80·95	86·25	151·1
·8458	80·90	86·25	151·0
·8459	80·85	86·20	150·9
·8460	80·80	86·15	150·9
·8461	80·80	86·15	150·8
·8462	80·75	86·10	150·7
·8463	80·70	86·05	150·7
·8464	80·65	86·05	150·6
·8465	80·65	86·00	150·6
·8466	80·60	86·00	150·5
·8467	80·55	85·95	150·5
·8468	80·50	85·95	150·4
·8469	80·50	85·90	150·3
·8470	80·45	85·90	150·3
·8471	80·40	85·85	150·2
·8472	80·35	85·80	150·2
·8473	80·30	85·80	150·1
·8474	80·30	85·75	150·1
·8475	80·25	85·70	150·0

I. Sp. gr. at 60° F. Water at 60°=1.	II. Percentage of alcohol *by weight.*	III. Percentage of alcohol *by volume.*	IV. Percentage of Proof Spirit.
·8476	80·20	85·70	150·0
·8477	80·15	85·65	149·9
·8478	80·10	85·60	149·9
·8479	80·10	85·55	149·8
·8480	80·05	85·50	149·7
·8481	80·00	85·50	149·7
·8482	79·95	85·45	149·6
·8483	79·90	85·40	149·5
·8484	79·85	85·35	149·5
·8485	79·80	85·35	149·4
·8486	79·75	85·30	149·3
·8487	79·70	85·25	149·3
·8488	79·65	85·25	149·2
·8489	79·60	85·20	149·2
·8490	79·60	85·15	149·1
·8491	79·55	85·10	149·1
·8492	79·50	85·05	149·0
·8493	79·45	85·00	149·0
·8494	79·40	85·00	148·9
·8495	79·40	84·95	148·9
·8496	79·35	84·90	148·8
·8497	79·30	84·90	148·8
·8498	79·30	84·85	148·7
·8499	79·25	84·80	148·6
·8500	79·20	84·80	148·6
·8501	79·15	84·75	148·5
·8502	79·10	84·70	148·5
·8503	79·05	84·70	148·4

I. Sp. gr. at 60° F. Water at 60°=1.	II. Percentage of alcohol *by weight*.	III. Percentage of alcohol *by volume*.	IV. Percentage of Proof Spirit.
·8504	79·00	84·65	148·4
·8505	79·00	84·60	148·3
·8506	78·95	84·60	148·2
·8507	78·90	84·55	148·2
·8508	78·90	84·50	148·1
·8509	78·85	84·50	148·1
·8510	78·80	84·45	148·0
·8511	78·75	84·45	147·9
·8512	78·70	84·40	147·9
·8513	78·70	84·35	147·8
·8514	78·65	84·30	147·7
·8515	78·60	84·25	147·7
·8516	78·55	84·25	147·6
·8517	78·50	84·20	147·6
·8518	78·50	84·20	147·5
·8519	78·45	84·15	147·4
·8520	78·40	84·15	147·4
·8521	78·40	84·10	147·3
·8522	78·35	84·05	147·3
·8523	78·30	84·05	147·2
·8524	78·25	84·00	147·1
·8525	78·20	83·95	147·1
·8526	78·15	83·90	147·0
·8527	78·10	83·90	146·9
·8528	78·05	83·85	146·8
·8529	78·00	83·85	146·8
·8530	78·00	83·80	146·7
·8531	77·95	83·80	146·7

I. Sp. gr. at 60° F. Water at 60°=1.	II. Percentage of alcohol *by weight*.	III. Percentage of alcohol *by volume*.	IV. Percentage of Proof Spirit.
·8532	77·90	83·75	146·6
·8533	77·85	83·75	146·6
·8534	77·80	83·70	146·5
·8535	77·75	83·65	146·5
·8536	77·70	83·65	146·4
·8537	77·70	83·60	146·4
·8538	77·65	83·55	146·3
·8539	77·60	83·50	146·2
·8540	77·55	83·45	146·2
·8541	77·50	83·45	146·1
·8542	77·50	83·40	146·1
·8543	77·45	83·35	146·0
·8544	77·45	83·30	146·0
·8545	77·40	83·25	145·9
·8546	77·35	83·25	145·8
·8547	77·30	83·20	145·8
·8548	77·25	83·20	145·7
·8549	77·20	83·15	145·6
·8550	77·15	83·15	145·6
·8551	77·10	83·10	145·5
·8552	77·05	83·05	145·4
·8553	77·00	83·00	145·4
·8554	77·00	82·95	145·3
·8555	76·95	82·90	145·3
·8556	76·90	82·85	145·2
·8557	76·85	82·80	145·2
·8558	76·80	82·80	145·1
·8559	76·75	82·75	145·0
·8560	76·70	82·75	145·0

I. Sp. gr. at 60° F. Water at 60°=1.	II. Percentage of alcohol *by* *weight*.	III. Percentage of alcohol *by* *volume*.	IV. Percentage of Proof Spirit.
·8561	76·70	82·70	144·9
·8562	76·65	82·65	144·8
·8563	76·60	82·65	144·8
·8564	76·55	82·60	144·7
·8565	76·50	82·60	144·6
·8566	76·50	82·50	144·6
·8567	76·45	82·50	144·5
·8568	76·40	82·45	144·5
·8569	76·35	82·40	144·4
·8570	76·30	82·40	144·4
·8571	76·25	82·35	144·3
·8572	76·20	82·30	144·3
·8573	76·20	82·30	144·2
·8574	76·15	82·25	144·2
·8575	76·10	82·20	144·1
·8576	76·05	82·20	144·1
·8577	76·00	82·15	144·0
·8578	75·95	82·10	143·9
·8579	75·90	82·05	143·9
·8580	75·85	82·00	143·8
·8581	75·80	81·95	143·7
·8582	75·80	81·95	143·7
·8583	75·75	81·90	143·6
·8584	75·70	81·85	143·6
·8585	75·70	81·85	143·5
·8586	75·65	81·80	143·4
·8587	75·60	81·80	143·4
·8588	75·60	81·75	143·3
·8589	75·55	81·70	143·3

I. Sp. gr. at 60° F. Water at 60°=1.	II. Percentage of alcohol *by weight*.	III. Percentage of alcohol *by volume*.	IV. Percentage of Proof Spirit.
·8590	75·50	81·70	143·2
·8591	75·45	81·65	143·1
·8592	75·45	81·60	143·0
·8593	75·40	81·60	143·0
·8594	75·35	81·55	142·9
·8595	75·30	81·50	142·9
·8596	75·25	81·50	142·8
·8597	75·20	81·45	142·8
·8598	75·20	81·40	142·7
·8599	75·15	81·40	142·7
·8600	75·10	81·35	142·6
·8601	75·05	81·30	142·5
·8602	75·00	81·30	142·4
·8603	75·00	81·25	142·4
·8604	74·95	81·20	142·3
·8605	74·90	81·20	142·3
·8606	74·85	81·15	142·2
·8607	74·85	81·15	142·2
·8608	74·80	81·10	142·1
·8609	74·75	81·05	142·0
·8610	74·70	81·00	141·9
·8611	74·65	80·95	141·9
·8612	74·60	80·95	141·8
·8613	74·55	80·90	141·7
·8614	74·50	80·85	141·7
·8615	74·50	80·80	141·6
·8616	74·45	80·75	141·5
·8617	74·40	80·75	141·5

I. Sp. gr. at 60° F. Water at 60°=1.	II. Percentage of alcohol *by* *weight*.	III. Percentage of alcohol *by* *volume*.	IV. Percentage of Proof Spirit.
·8618	74·35	80·70	141·4
·8619	74·30	80·65	141·3
·8620	74·25	80·60	141·3
·8621	74·20	80·55	141·2
·8622	74·20	80·55	141·2
·8623	74·15	80·50	141·1
·8624	74·10	80·45	141·0
·8625	74·05	80·40	140·9
·8626	74·00	80·40	140·9
·8627	73·95	80·35	140·8
·8628	73·90	80·30	140·8
·8629	73·85	80·25	140·7
·8630	73·80	80·25	140·7
·8631	73·75	80·20	140·6
·8632	73·75	80·20	140·5
·8633	73·70	80·15	140·5
·8634	73·65	80·10	140·4
·8635	73·60	80·10	140·3
·8636	73·60	80·05	140·3
·8637	73·55	80·05	140·2
·8638	73·50	80·00	140·2
·8639	73·50	79·95	140·1
·8640	73·45	79·95	140·1
·8641	73·40	79·90	140·0
·8642	73·35	79·85	139·9
·8643	73·30	79·80	139·9
·8644	73·25	79·80	139·8
·8645	73·20	79·75	139·7

I. Sp. gr. at 60° F. Water at 60°=1.	II. Percentage of alcohol *by weight.*	III. Percentage of alcohol *by volume.*	IV. Percentage of Proof Spirit.
·8646	73·20	79·70	139·7
·8647	73·15	79·65	139·6
·8648	73·10	79·65	139·5
·8649	73·05	79·60	139·5
·8650	73·00	79·55	139·4
·8651	73·00	79·55	139·4
·8652	72·95	79·50	139·3
·8653	72·90	79·45	139·2
·8654	72·85	79·40	139·1
·8655	72·80	79·35	139·1
·8656	72·75	79·30	139·0
·8657	72·70	79·30	139·0
·8658	72·65	79·25	138·9
·8659	72·60	79·20	138·8
·8660	72·55	79·15	138·8
·8661	72·50	79·10	138·7
·8662	72·50	79·10	138·6
·8663	72·45	79·05	138·6
·8664	72·40	79·00	138·5
·8665	72·40	78·95	138·4
·8666	72·35	78·95	138·3
·8667	72·30	78·90	138·3
·8668	72·30	78·85	138·2
·8669	72·25	78·80	138·1
·8670	72·20	78·75	138·1
·8671	72·15	78·75	138·0
·8672	72·10	78·70	137·9
·8673	72·05	78·70	137·9

I. Sp. gr. at 60° F. Water at 60°=1.	II. Percentage of alcohol *by* *weight*.	III. Percentage of alcohol *by* *volume*.	IV. Percentage of Proof Spirit.
·8674	72·00	78·65	137·8
·8675	71·95	78·60	137·7
·8676	71·90	78·60	137·7
·8677	71·85	78·55	137·6
·8678	71·80	78·50	137·5
·8679	71·80	78·50	137·5
·8680	71·75	78·45	137·4
·8681	71·70	78·40	137·3
·8682	71·65	78·35	137·2
·8683	71·60	78·35	137·2
·8684	71·55	78·30	137·1
·8685	71·50	78·25	137·0
·8686	71·45	78·25	136·9
·8687	71·40	78·20	136·8
·8688	71·40	78·15	136·8
·8689	71·35	78·10	136·7
·8690	71·30	78·10	136·6
·8691	71·25	78·05	136·6
·8692	71·25	78·00	ʻ136·5
·8693	71·20	77·95	136·5
·8694	71·15	77·95	136·4
·8695	71·10	77·90	136·4
·8696	71·05	77·90	136·3
·8697	71·00	77·85	136·3
·8698	70·95	77·80	136·2
·8699	70·90	77·80	136·2
·8700	70·85	77·70	136·1
·8701	70·85	77·65	136·1

I.	II.	III.	IV.
Sp. gr. at 60° F. Water at 60°=1.	Percentage of alcohol *by weight*.	Percentage of alcohol *by volume*.	Percentage of Proof Spirit.
·8702	70·80	77·65	136·0
·8703	70·75	77·60	136·0
·8704	70·75	77·60	135·9
·8705	70·70	77·55	135·9
·8706	70·70	77·50	135·8
·8707	70·65	77·50	135·7
·8708	70·60	77·45	135·7
·8709	70·55	77·40	135·6
·8710	70·50	77·35	135·6
·8711	70·45	77·30	135·5
·8712	70·40	77·30	135·4
·8713	70·35	77·25	135·4
·8714	70·30	77·20	135·3
·8715	70·30	77·20	135·2
·8716	70·25	77·15	135·2
·8717	70·20	77·10	135·1
·8718	70·15	77·05	135·0
·8719	70·10	77·00	135·0
·8720	70·05	76·95	134·9
·8721	70·00	76·95	134·8
·8722	70·00	76·90	134·8
·8723	69·95	76·85	134·7
·8724	69·90	76·80	134·6
·8725	69·85	76·80	134·6
·8726	69·80	76·75	134·5
·8727	69·80	76·70	134·5
·8728	69·75	76·65	134·4
·8729	69·70	76·65	134·3
·8730	69·65	76·60	134·3

I. Sp. gr. at 60° F. Water at 60°=1.	II. Percentage of alcohol *by weight*.	III. Percentage of alcohol *by volume*.	IV. Percentage of Proof Spirit.
·8731	69·60	76·55	134·2
·8732	69·55	76·55	134·2
·8733	69·50	76·50	134·1
·8734	69·45	76·40	134·0
·8735	69·40	76·35	133·9
·8736	69·40	76·30	133·8
·8737	69·35	76·25	133·8
·8738	69·30	76·25	133·7
·8739	69·25	76·20	133·6
·8740	69·20	76·15	133·5
·8741	69·15	76·15	133·5
·8742	69·10	76·10	133·4
·8743	69·05	76·05	133·3
·8744	69·00	76·00	133·2
·8745	69·00	76·00	133·1
·8746	68·95	75·95	133·1
·8747	68·90	75·90	133·0
·8748	68·85	75·85	133·0
·8749	68·80	75·80	132·9
·8750	68·80	75·80	132·9
·8751	68·75	75·75	132·8
·8752	68·70	75·70	132·7
·8753	68·65	75·70	132·7
·8754	68·60	75·65	132·6
·8755	68·60	75·60	132·5
·8756	68·55	75·60	132·5
·8757	68·50	75·55	132·4
·8758	68·50	75·55	132·4
·8759	68·45	75·50	132·3

I. Sp. gr. at 60° F. Water at 60°=1.	II. Percentage of alcohol *by weight.*	III. Percentage of alcohol *by volume.*	IV. Percentage of Proof Spirit.
·8760	68·40	75·45	132·2
·8761	68·35	75·40	132·2
·8762	68·30	75·40	132·1
·8763	68·25	75·30	132·0
·8764	68·20	75·25	132·0
·8765	68·20	75·25	131·9
·8766	68·15	75·20	131·8
·8767	68·10	75·15	131·8
·8768	68·05	75·15	131·7
·8769	68·00	75·10	131·6
·8770	67·95	75·10	131·6
·8771	67·90	75·05	131·5
·8772	67·90	75·00	131·4
·8773	67·85	74·95	131·4
·8774	67·80	74·90	131·3
·8775	67·75	74·90	131·2
·8776	67·75	74·85	131·2
·8777	67·70	74·85	131·1
·8778	67·65	74·80	131·0
·8779	67·60	74·75	131·0
·8780	67·55	74·70	130·9
·8781	67·50	74·65	130·8
·8782	67·45	74·65	130·8
·8783	67·40	74·60	130·7
·8784	67·35	74·55	130·6
·8785	67·30	74·50	130·6
·8786	67·25	74·45	130·5
·8787	67·20	74·40	130·4

I. Sp. gr. at 60° F. Water at 60°=1.	II. Percentage of alcohol *by weight*.	III. Percentage of alcohol *by volume*.	IV. Percentage of Proof Spirit.
·8788	67·15	74·40	130·4
·8789	67·10	74·35	130·3
·8790	67·05	74·30	130·2
·8791	67·00	74·25	130·2
·8792	67·00	74·20	130·1
·8793	66·95	74·15	130·0
·8794	66·90	74·15	130·0
·8795	66·85	74·10	129·9
·8796	66·80	74·05	129·8
·8797	66·75	74·00	129·8
·8798	66·75	74·00	129·7
·8799	66·70	73·95	129·6
·8800	66·65	73·90	129·6
·8801	66·60	73·85	129·5
·8802	66·55	73·85	129·4
·8803	66·50	73·80	129·4
·8804	66·50	73·75	129·3
·8805	66·45	73·70	129·2
·8806	66·40	73·65	ʼ129·2
·8807	66·35	73·60	129·1
·8808	66·30	73·60	129·0
·8809	66·25	73·55	128·9
·8810	66·25	73·50	128·9
·8811	66·20	73·45	128·8
·8812	66·15	73·45	128·7
·8813	66·10	73·40	128·6
·8814	66·05	73·40	128·6
·8815	66·00	73·35	128·5

I. Sp. gr. at 60° F. Water at 60°=1.	II. Percentage of alcohol *by weight*.	III. Percentage of alcohol *by volume*.	IV. Percentage of Proof Spirit.
·8816	66·00	73·30	128·4
·8817	65·95	73·25	128·3
·8818	65·90	73·25	128·3
·8819	65·85	73·20	128·2
·8820	65·80	73·15	128·2
·8821	65·75	73·10	128·1
·8822	65·70	73·05	128·1
·8823	65·70	73·05	128·0
·8824	65·65	73·00	127·9
·8825	65·60	72·95	127·9
·8826	65·60	72·90	127·8
·8827	65·55	72·85	127·8
·8828	65·50	72·85	127·7
·8829	65·45	72·80	127·6
·8830	65·40	72·75	127·5
·8831	65·40	72·75	127·5
·8832	65·35	72·70	127·4
·8833	65·30	72·65	127·3
·8834	65·25	72·60	127·2
·8835	65·25	72·55	127·1
·8836	65·20	72·50	127·0
·8837	65·15	72·45	127·0
·8838	65·10	72·40	126·9
·8839	65·05	72·40	126·8
·8840	65·00	72·35	126·8
·8841	64·95	72·35	126·7
·8842	64·90	72·30	126·6
·8843	64·85	72·25	126·5

I. Sp. gr. at 60° F. Water at 60°=1.	II. Percentage of alcohol *by weight*.	III. Percentage of alcohol *by volume*.	IV. Percentage of Proof Spirit.
·8844	64·80	72·20	126·5
·8845	64·80	72·15	126·4
·8846	64·75	72·10	126·3
·8847	64·70	72·05	126·3
·8848	64·65	72·00	126·2
·8849	64·60	71·95	126·1
·8850	64·55	71·90	126·0
·8851	64·50	71·90	126·0
·8852	64·45	71·85	125·9
·8853	64·40	71·80	125·8
·8854	64·35	71·80	125·7
·8855	64·35	71·75	125·7
·8856	64·30	71·70	125·6
·8857	64·25	71·70	125·6
·8858	64·20	71·65	125·5
·8859	64·20	71·60	125·5
·8860	64·15	71·60	125·4
·8861	64·10	71·55	125·4
·8862	64·05	71·50	125·3
·8863	64·00	71·45	125·3
·8864	63·95	71·40	125·2
·8865	63·90	71·40	125·1
·8866	63·90	71·35	125·0
·8867	63·85	71·30	125·0
·8868	63·80	71·25	124·9
·8869	63·75	71·20	124·8
·8870	63·70	71·20	124·7
·8871	63·65	71·15	124·6

I. Sp. gr. at 60° F. Water at 60°=1.	II. Percentage of alcohol *by weight*.	III. Percentage of alcohol *by volume*.	IV. Percentage of Proof Spirit.
·8872	63·60	71·10	124·6
·8873	63·55	71·05	124·5
·8874	63·50	71·00	124·4
·8875	63·45	71·00	124·3
·8876	63·40	70·95	124·3
·8877	63·35	70·90	124·2
·8878	63·30	70·85	124·1
·8879	63·30	70·80	124·0
·8880	63·25	70·75	124·0
·8881	63·20	70·70	123·9
·8882	63·15	70·65	123·8
·8883	63·15	70·65	123·8
·8884	63·10	70·60	123·7
·8885	63·05	70·55	123·7
·8886	63·00	70·55	123·6
·8887	63·00	70·50	123·5
·8888	62·95	70·45	123·5
·8889	62·90	70·40	123·4
·8890	62·85	70·35	123·3
·8891	62·80	70·35	123·3
·8892	62·75	70·30	123·2
·8893	62·70	70·25	123·1
·8894	62·65	70·20	123·1
·8895	62·60	70·20	123·0
·8896	62·60	70·15	122·9
·8897	62·55	70·10	122·8
·8898	62·50	70·05	122·8
·8899	62·50	70·00	122·7
·8900	62·45	69·95	122·6

I. Sp. gr. at 60° F. Water at 60°=1.	II. Percentage of alcohol *by weight*.	III. Percentage of alcohol *by volume*.	IV. Percentage of Proof Spirit.
·8901	62·40	69·95	122·5
·8902	62·35	69·90	122·5
·8903	62·30	69·85	122·4
·8904	62·25	69·80	122·3
·8905	62·20	69·80	122·3
·8906	62·15	69·75	122·2
·8907	62·10	69·70	122·1
·8908	62·05	69·65	122·1
·8909	62·00	69·60	122·0
·8910	62·00	69·60	121·9
·8911	61·95	69·55	121·8
·8912	61·90	69·50	121·8
·8913	61·85	69·45	121·7
·8914	61·80	69·40	121·7
·8915	61·75	69·35	121·6
·8916	61·70	69·30	121·5
·8917	61·70	69·25	121·4
·8918	61·65	69·25	121·3
·8919	61·60	69·20	121·2
·8920	61·55	69·15	121·1
·8921	61·50	69·10	121·1
·8922	61·45	69·05	121·0
·8923	61·40	69·00	120·9
·8924	61·35	69·00	120·9
·8925	61·30	68·95	120·8
·8926	61·30	68·90	120·7
·8927	61·25	68·85	120·7
·8928	61·20	68·85	120·6
·8929	61·15	68·80	120·5

I. Sp. gr. at 60° F. Water at 60°=1.	II. Percentage of alcohol *by weight*.	III. Percentage of alcohol *by volume*.	IV. Percentage of Proof Spirit.
·8930	61·10	68·75	120·5
·8931	61·10	68·70	120·5
·8932	61·05	68·70	120·4
·8933	61·00	68·65	120·3
·8934	60·95	68·60	120·2
·8935	60·90	68·60	120·2
·8936	60·90	68·55	120·1
·8937	60·85	68·50	120·0
·8938	60·80	68·45	120·0
·8939	60·75	68·40	119·9
·8940	60·70	68·35	119·8
·8941	60·70	68·35	119·8
·8942	60·65	68·30	119·7
·8943	60·60	68·25	119·6
·8944	60·55	68·20	119·6
·8945	60·50	68·20	119·5
·8946	60·45	68·15	119·4
·8947	60·40	68·10	119·4
·8948	60·35	68·05	119·3
·8949	60·30	68·00	119·2
·8950	60·30	68·00	119·2
·8951	60·25	67·95	119·1
·8952	60·20	67·90	119·0
·8953	60·15	67·85	118·9
·8954	60·10	67·80	118·9
·8955	60·05	67·80	118·8
·8956	60·00	67·75	118·7
·8957	60·00	67·70	118·6

I. Sp. gr. at 60° F. Water at 60°=1.	II. Percentage of alcohol *by weight.*	III. Percentage of alcohol *by volume.*	IV. Percentage of Proof Spirit.
·8958	59·95	67·65	118·5
·8959	59·90	67·60	118·5
·8960	59·85	67·55	118·4
·8961	59·80	67·50	118·3
·8962	59·80	67·50	118·3
·8963	59·75	67·45	118·2
·8964	59·70	67·40	118·1
·8965	59·65	67·35	118·0
·8966	59·60	67·25	117·9
·8967	59·50	67·20	117·8
·8968	59·45	67·15	117·7
·8969	59·40	67·10	117·6
·8970	59·35	67·05	117·5
·8971	59·30	67·00	117·4
·8972	59·25	66·95	117·4
·8973	59·20	66·90	117·3
·8974	59·15	66·85	117·2
·8975	59·15	66·80	117·1
·8976	59·10	66·75	117·0
·8977	59·05	66·70	117·0
·8978	59·00	66·70	116·9
·8979	59·00	66·65	116·9
·8980	58·95	66·65	116·8
·8981	58·90	66·60	116·7
·8982	58·85	66·60	116·7
·8983	58·80	66·55	116·6
·8984	58·80	66·50	116·6
·8985	58·75	66·50	116·5

I. Sp. gr. at 60° F. Water at 60°=1.	II. Percentage of alcohol *by weight.*	III. Percentage of alcohol *by volume.*	IV. Percentage of Proof Spirit.
·8986	58·70	66·45	116·5
·8987	58·70	66·40	116·4
·8988	58·65	66·40	116·4
·8989	58·60	66·35	116·3
·8990	58·55	66·30	116·2
·8991	58·50	66·25	116·1
·8992	58·45	66·20	116·0
·8993	58·40	66·15	116·0
·8994	58·35	66·10	115·9
·8995	58·30	66·05	115·8
·8996	58·25	66·00	115·7
·8997	58·20	65·95	115·6
·8998	58·15	65·90	115·5
·8999	58·10	65·85	115·5
·9000	58·05	65·80	115·4
·9001	58·00	65·75	115·3
·9002	58·00	65·70	115·2
·9003	57·95	65·70	115·1
·9004	57·90	65·65	115·0
·9005	57·85	65·60	115·0
·9006	57·80	65·55	114·9
·9007	57·75	65·50	114·8
·9008	57·70	65·45	114·7
·9009	57·65	65·40	114·6
·9010	57·60	65·35	114·6
·9011	57·55	65·30	114·5
·9012	57·50	65·30	114·4
·9013	57·50	65·25	114·4

I. Sp. gr. at 60° F. Water at 60°=1.	II. Percentage of alcohol *by* *weight*.	III. Percentage of alcohol *by* *volume*.	IV. Percentage of Proof Spirit.
·9014	57·45	65·20	114·3
·9015	57·40	65·15	114·2
·9016	57·35	65·10	114·1
·9017	57·35	5·10	114·1
·9018	57·30	65·05	114·0
·9019	57·25	65·00	114·0
·9020	57·20	65·00	113·9
·9021	57·20	64·95	113·8
·9022	57·15	64·90	113·8
·9023	57·10	64·90	113·7
·9024	57·05	64·85	113·6
·9025	57·00	64·80	113·5
·9026	56·95	64·75	113·4
·9027	56·90	64·70	113·3
·9028	56·85	64·65	113·2
·9029	56·80	64·60	113·1
·9030	56·75	64·55	113·1
·9031	56·70	64·50	113·0
·9032	56·70	64·45	·112·9
·9033	56·65	64·40	112·8
·9034	56·60	64·35	112·8
·9035	56·55	64·30	112·7
·9036	56·50	64·25	112·6
·9037	56·45	64·20	112·5
·9038	56·40	64·20	112·5
·9039	56·35	64·15	112·4
·9040	56·35	64·10	112·3
·9041	56·30	64·05	112·2

I. Sp. gr. at 60° F. Water at 60° = 1.	II. Percentage of alcohol *by* *weight.*	III. Percentage of alcohol *by* *volume.*	IV. Percentage of Proof Spirit.
·9042	56·25	64·00	112·2
·9043	56·20	64·00	112·1
·9044	56·15	63·95	112·0
·9045	56·10	63·90	112·0
·9046	56·05	63·85	111·9
·9047	56·00	63·80	111·8
·9048	56·00	63·80	111·8
·9049	55·95	63·75	111·7
·9050	55·90	63·70	111·6
·9051	55·85	63·70	111·6
·9052	55·80	63·65	111·5
·9053	55·75	63·60	111·4
·9054	55·75	63·60	111·4
·9055	55·70	63·55	111·3
·9056	55·65	63·50	111·2
·9057	55·60	63·45	111·1
·9058	55·55	63·40	111·0
·9059	55·50	63·35	111·0
·9060	55·45	63·30	110·9
·9061	55·40	63·25	110·8
·9062	55·35	63·20	110·7
·9063	55·30	63·15	110·6
·9064	55·25	63·10	110·5
·9065	55·20	63·05	110·4
·9066	55·15	63·00	110·3
·9067	55·10	62·95	110·2
·9068	55·05	62·90	110·2
·9069	55·00	62·85	110·1
·9070	54·95	62·80	110·0

I. Sp. gr. at 60° F. Water at 60°=1.	II. Percentage of alcohol *by weight.*	III. Percentage of alcohol *by volume.*	IV. Percentage of Proof Spirit.
·9071	54·95	62·75	110·0
·9072	54·90	62·75	109·9
·9073	54·85	62·70	109·8
·9074	54·85	62·65	109·7
·9075	54·80	62·60	109·6
·9076	54·75	62·55	109·6
·9077	54·70	62·55	109·5
·9078	54·65	62·50	109·5
·9079	54·60	62·45	109·4
·9080	54·55	62·40	109·3
·9081	54·50	62·35	109·2
·9082	54·45	62·30	109·2
·9083	54·40	62·25	109·1
·9084	54·35	62·20	109·0
·9085	54·30	62·15	108·9
·9086	54·30	62·15	108·9
·9087	54·25	62·10	108·8
·9088	54·20	62·05	108·7
·9089	54·15	62·00	108·6
·9090	54·10	61·95	108·5
·9091	54·10	61·90	108·5
·9092	54·05	61·85	108·4
·9093	54·00	61·80	108·3
·9094	53·95	61·75	108·3
·9095	53·90	61·70	108·2
·9096	53·85	61·65	108·1
·9097	53·80	61·60	108·0
·9098	53·75	61·60	108·0
·9099	53·70	61·55	107·9

I. Sp. gr. at 60° F. Water at 60°=1.	II. Percentage of alcohol *by weight*.	III. Percentage of alcohol *by volume*.	IV. Percentage of Proof Spirit.
·9100	53·65	61·50	107·8
·9101	53·65	61·45	107·7
·9102	53·60	61·40	107·7
·9103	53·55	61·40	107·6
·9104	53·50	61·35	107·5
·9105	53·45	61·30	107·4
·9106	53·40	61·25	107·3
·9107	53·35	61·20	107·3
·9108	53·35	61·15	107·2
·9109	53·30	61·10	107·1
·9110	53·25	61·05	107·0
·9111	53·20	61·00	106·9
·9112	53·15	60·95	106·9
·9113	53·10	60·95	106·8
·9114	53·05	60·90	106·7
·9115	53·00	60·85	106·6
·9116	52·95	60·80	106·6
·9117	52·90	60·75	106·5
·9118	52·90	60·75	106·4
·9119	52·85	60·70	106·4
·9120	52·80	60·65	106·3
·9121	52·75	60·60	106·2
·9122	52·70	60·55	106·1
·9123	52·70	60·50	106·0
·9124	52·65	60·45	105·9
·9125	52·60	60·40	105·9
·9126	52·55	60·35	105·8
·9127	52·50	60·30	105·7

I. Sp. gr. at 60° F. Water at 60°=1.	II. Percentage of alcohol *by weight*.	III. Percentage of alcohol *by volume*.	IV. Percentage of Proof Spirit.
·9128	52·45	60·25	105·7
·9129	52·40	60·20	105·6
·9130	52·35	60·15	105·5
·9131	52·30	60·15	105·4
·9132	52·25	60·10	105·4
·9133	52·25	60·05	105·3
·9134	52·20	60·00	105·2
·9135	52·15	60·00	105·2
·9136	52·10	59·95	105·1
·9137	52·05	59·90	105·0
·9138	52·00	59·85	104·9
·9139	51·95	59·80	104·8
·9140	51·90	59·75	104·7
·9141	51·85	59·70	104·6
·9142	51·80	59·65	104·5
·9143	51·75	59·60	104·4
·9144	51·70	59·55	104·4
·9145	51·65	59·50	104·3
·9146	51·65	59·45	104·2
·9147	51·60	59·40	104·2
·9148	51·55	59·40	104·1
·9149	51·50	59·35	104·0
·9150	51·45	59·30	103·9
·9151	51·40	59·25	103·9
·9152	51·35	59·25	103·8
·9153	51·35	59·20	103·8
·9154	51·30	59·15	103·7
·9155	51·25	59·10	103·6

I. Sp. gr. at 60° F. Water at 60°=1.	II. Percentage of alcohol *by* *weight*.	III. Percentage of alcohol *by* *volume*.	IV. Percentage of Proof Spirit.
·9156	51·20	59·05	103·5
·9157	51·15	59·00	103·4
·9158	51·10	58·95	103·3
·9159	51·05	58·90	103·2
·9160	51·00	58·85	103·1
·9161	50·95	58·80	103·0
·9162	50·90	58·75	102·9
·9163	50·85	58·70	102·9
·9164	50·80	58·65	102·8
·9165	50·75	58·60	102·7
·9166	50·75	58·55	102·6
·9167	50·70	58·50	102·5
·9168	50·65	58·50	102·5
·9169	50·60	58·45	102·4
·9170	50·55	58·40	102·3
·9171	50·50	58·35	102·2
·9172	50·45	58·30	102·1
·9173	50·40	58·25	102·0
·9174	50·35	58·20	102·0
·9175	50·30	58·15	101·9
·9176	50·30	58·10	101·8
·9177	50·25	58·05	101·7
·9178	50·20	58·00	101·7
·9179	50·15	57·95	101·6
·9180	50·10	57·90	101·5
·9181	50·05	57·85	101·4
·9182	50·00	57·80	101·3
·9183	49·95	57·75	101·2

I. Sp. gr. at 60° F. Water at 60°=1.	II. Percentage of alcohol *by weight.*	III. Percentage of alcohol *by volume.*	IV. Percentage of Proof Spirit.
·9184	49·90	57·70	101·1
·9185	49·85	57·65	101·0
·9186	49·80	57·60	100·9
·9187	49·75	57·55	100·8
·9188	49·75	57·50	100·7
·9189	49·70	57·45	100·6
·9190	49·65	57·40	100·6
·9191	49·60	57·40	100·5
·9192	49·55	57·35	100·4
·9193	49·50	57·30	100·4
·9194	49·45	57·25	100·3
·9195	49·40	57·20	100·2
·9196	49·35	57·15	100·2
·9197	49·30	57·10	100·1
·9198	49·25	57·05	100·0
·9199	49·20	57·00	99·9
·9200	49·15	56·95	99·8
·9201	49·10	56·90	99·7
·9202	49·00	56·85	99·6
·9203	48·95	56·80	99·5
·9204	48·90	56·70	99·4
·9205	48·85	56·65	99·3
·9206	48·80	56·60	99·2
·9207	48·80	56·60	99·2
·9208	48·75	56·55	99·1
·9209	48·70	56·50	99·0
·9210	48·65	56·50	99·0
·9211	48·65	56·45	98·9

I. Sp. gr. at 60° F. Water at 60°=1.	II. Percentage of alcohol *by weight*.	III. Percentage of alcohol *by volume*.	IV. Percentage of Proof Spirit.
·9212	48·60	56·40	98·9
·9213	48·55	56·35	98·8
·9214	48·50	56·30	98·7
·9215	48·50	56·30	98·6
·9216	48·45	56·25	98·6
·9217	48·40	56·20	98·5
·9218	48·35	56·15	98·4
·9219	48·30	56·10	98·3
·9220	48·25	56·05	98·2
·9221	48·20	56·00	98·1
·9222	48·15	55·95	98·0
·9223	48·10	55·90	98·0
·9224	48·05	55·85	97·9
·9225	48·00	55·80	97·8
·9226	48·00	55·75	97·7
·9227	47·95	55·70	97·6
·9228	47·90	55·65	97·5
·9229	47·85	55·60	97·5
·9230	47·80	55·55	97·4
·9231	47·75	55·50	97·3
·9232	47·70	55·45	97·2
·9233	47·65	55·40	97·1
·9234	47·60	55·35	97·0
·9235	47·55	55·30	97·0
·9236	47·50	55·30	96·9
·9237	47·45	55·25	96·8
·9238	47·40	55·20	96·7
·9239	47·35	55·15	96·6
·9240	47·30	55·10	96·5

I. Sp gr. at 60° F. Water at 60°=1.	II. Percentage of alcohol *by weight.*	III. Percentage of alcohol *by volume.*	IV. Percentage of Proof Spirit.
·9241	47·25	55·05	96·4
·9242	47·25	55·00	96·3
·9243	47·20	54·95	96·2
·9244	47·15	54·90	96·1
·9245	47·10	54·85	96·0
·9246	47·05	54·80	95·9
·9247	47·00	54·75	95·8
·9248	47·00	54·70	95·7
·9249	46·95	54·65	95·7
·9250	46·90	54·60	95·6
·9251	46·85	54·55	95·5
·9252	46·80	54·50	95·4
·9253	46·75	54·45	95·3
·9254	46·70	54·40	95·3
·9255	46·65	54·35	95·2
·9256	46·60	54·30	95·1
·9257	46·55	54·25	95·0
·9258	46·50	54·20	95·0
·9259	46·45	54·15	94·9
·9260	46·40	54·10	94·8
·9261	46·35	54·10	94·8
·9262	46·30	54·05	94·7
·9263	46·25	54·00	94·6
·9264	46·25	53·95	94·6
·9265	46·20	53·90	94·5
·9266	46·15	53·85	94·4
·9267	46·10	53·80	94·4
·9268	46·05	53·75	94·3
·9269	46·00	53·70	94·2

I. Sp. gr. at 60° F. Water at 60°=1.	II. Percentage of alcohol *by weight*.	III. Percentage of alcohol *by volume*.	IV. Percentage of Proof Spirit.
·9270	45·95	53·65	94·1
·9271	45·90	53·60	94·0
·9272	45·85	53·60	94·0
·9273	45·80	53·55	93·9
·9274	45·80	53·50	93·8
·9275	45·75	53·45	93·7
·9276	45·70	53·40	93·6
·9277	45·65	53·35	93·5
·9278	45·60	53·30	93·4
·9279	45·55	53·20	93·3
·9280	45·50	53·15	93·2
·9281	45·45	53·10	93·1
·9282	45·40	53·05	93·0
·9283	45·35	53·00	92·9
·9284	45·30	52·95	92·8
·9285	45·25	52·90	92·7
·9286	45·20	52·85	92·6
·9287	45·15	52·85	92·6
·9288	45·10	52·80	92·5
·9289	45·05	52·75	92·4
·9290	45·00	52·70	92·3
·9291	44·95	52·60	92·2
·9292	44·90	52·55	92·1
·9293	44·85	52·50	92·0
·9294	44·80	52·45	91·9
·9295	44·75	52·40	91·8
·9296	44·75	52·35	91·8
·9297	44·70	52·30	91·7

I. Sp. gr. at 60° F. Water at 60°=1.	II. Percentage of alcohol *by weight*.	III. Percentage of alcohol *by volume*.	IV. Percentage of Proof Spirit.
·9298	44·65	52·25	91·6
·9299	44·60	52·20	91·5
·9300	44·55	52·15	91·4
·9301	44·50	52·10	91·3
·9302	44·45	52·05	91·2
·9303	44·40	52·00	91·2
·9304	44·35	51·95	91·1
·9305	44·30	51·90	91·0
·9306	44·25	51·85	91·0
·9307	44·20	51·80	90·9
·9308	44·20	51·80	90·8
·9309	44·15	51·75	90·7
·9310	44·10	51·70	90·6
·9311	44·05	51·65	90·6
·9312	44·00	51·60	90·5
·9313	43·95	51·55	90·4
·9314	43·90	51·50	90·3
·9315	43·85	51·45	90·2
·9316	43·80	51·40	90·2
·9317	43·75	51·35	90·1
·9318	43·70	51·30	90·0
·9319	43·65	51·25	89·9
·9320	43·60	51·20	89·8
·9321	43·55	51·15	89·7
·9322	43·50	51·10	89·6
·9323	43·50	51·05	89·5
·9324	43·45	51·00	89·4
·9325	43·40	50·95	89·3

I. Sp. gr. at 60° F. Water at 60°=1.	II. Percentage of alcohol *by weight*.	III. Percentage of alcohol *by volume*.	IV. Percentage of Proof Spirit.
·9326	43·35	50·90	89·2
·9327	43·30	50·85	89·1
·9328	43·25	50·80	89·0
·9329	43·20	50·75	89·0
·9330	43·15	50·70	88·9
·9331	43·10	50·65	88·8
·9332	43·05	50·60	88·7
·9333	43·00	50·55	88·6
·9334	42·90	50·50	88·5
·9335	42·90	50·45	88·4
·9336	42·85	50·40	88·3
·9337	42·80	50·30	88·2
·9338	42·75	50·25	88·1
·9339	42·65	50·20	88·0
·9340	42·60	50·15	87·9
·9341	42·55	50·10	87·8
·9342	42·55	50·05	87·7
·9343	42·50	50·00	87·6
·9344	42·45	49·95	87·5
·9345	42·40	49·90	87·4
·9346	42·35	49·85	87·3
·9347	42·30	49·80	87·2
·9348	42·25	49·75	87·1
·9349	42·20	49·70	87·1
·9350	42·15	49·65	87·0
·9351	42·10	49·60	· 86·9
·9352	42·05	49·50	86·8
·9353	42·00	49·45	86·7

I. Sp. gr. at 60° F. Water at 60°=1.	II. Percentage of alcohol *by weight*.	III. Percentage of alcohol *by volume*.	IV. Percentage of Proof Spirit.
·9354	41·95	49·40	86·6
·9355	41·90	49·35	86·5
·9356	41·85	49·30	86·4
·9357	41·80	49·25	86·3
·9358	41·75	49·20	86·2
·9359	41·70	49·15	86·1
·9360	41·65	49·10	86·1
·9361	41·60	49·05	86·0
·9362	41·55	49·00	85·9
·9363	41·50	48·95	85·8
·9364	41·45	48·90	85·7
·9365	41·40	48·85	85·6
·9366	41·35	48·80	85·5
·9367	41·30	48·75	85·4
·9368	41·25	48·70	85·3
·9369	41·20	48·65	85·2
·9370	41·15	48·60	85·1
·9371	41·10	48·55	85·0
·9372	41·05	48·50	84·9
·9373	41·00	48·45	84·8
·9374	40·95	48·40	84·7
·9375	40·90	48·30	84·6
·9376	40·85	48·25	84·5
·9377	40·80	48·20	84·4
·9378	40·75	48·15	84·3
·9379	40·70	48·10	84·2
·9380	40·65	48·05	84·1
·9381	40·60	48·00	84·0

E 2

I. Sp. gr. at 60° F. Water at 60°=1.	II. Percentage of alcohol *by weight*.	III. Percentage of alcohol *by volume*.	IV. Percentage of Proof Spirit.
·9382	40·55	47·95	84·0
·9383	40·50	47·90	83·9
·9384	40·45	47·85	83·8
·9385	40·40	47·80	83·7
·9386	40·40	47·75	83·7
·9387	40·35	47·70	83·6
·9388	40·25	47·65	83·5
·9389	40·20	47·60	83·4
·9390	40·15	47·50	83·3
·9391	40·10	47·45	83·2
·9392	40·05	47·40	83·1
·9393	40·00	47·35	83·0
·9394	39·95	47·30	82·9
·9395	39·90	47·25	82·8
·9396	39·85	47·20	82·7
·9397	39·85	47·15	82·6
·9398	39·80	47·10	82·6
·9399	39·75	47·05	82·5
·9400	39·70	47·00	82·4
·9401	39·65	46·95	82·3
·9402	39·60	46·90	82·2
·9403	39·55	46·85	82·1
·9404	39·50	46·80	82·0
·9405	39·40	46·75	81·9
·9406	39·35	46·65	81·8
·9407	39·30	46·60	81·7
·9408	39·25	46·55	81·6
·9409	39·20	46·50	81·5
·9410	39·15	46·40	81·4

I. Sp. gr. at 60° F. Water at 60°=1.	II. Percentage of alcohol *by weight.*	III. Percentage of alcohol *by volume.*	IV. Percentage of Proof Spirit.
·9411	39·10	46·35	81·3
·9412	39·05	46·30	81·2
·9413	39·00	46·25	81·1
·9414	38·95	46·20	81·0
·9415	38·90	46·10	80·9
·9416	38·85	46·05	80·8
·9417	38·80	46·00	80·7
·9418	38·75	45·95	80·6
·9419	38·70	45·90	80·5
·9420	38·65	45·85	80·4
·9421	38·60	45·80	80·3
·9422	38·55	45·75	80·2
·9423	38·50	45·70	80·1
·9424	38·45	45·65	80·0
·9425	38·45	45·65	80·0
·9426	38·40	45·60	79·9
·9427	38·35	45·55	79·8
·9428	38·30	45·50	79·7
·9429	38·25	45·45	79·6
·9430	38·20	45·40	79·5
·9431	38·15	45·35	79·4
·9432	38·10	45·25	79·3
·9433	38·00	45·20	79·2
·9434	37·95	45·10	79·0
·9435	37·90	45·05	78·9
·9436	37·85	45·00	78·8
·9437	37·80	44·95	78·7
·9438	37·75	44·90	78·6
·9439	37·70	44·80	78·5

I. Sp. gr. at 60° F. Water at 60°=1.	II. Percentage of alcohol *by weight*.	III. Percentage of alcohol *by volume*.	IV. Percentage of Proof Spirit.
·9440	37·65	44·75	78·4
·9441	37·60	44·70	78·3
·9442	37·55	44·65	78·2
·9443	37·50	44·55	78·1
·9444	37·40	44·50	78·0
·9445	37·35	44·45	77·9
·9446	37·30	44·35	77·8
·9447	37·25	44·30	77·7
·9448	37·20	44·25	77·6
·9449	37·15	44·20	77·5
·9450	37·10	44·15	77·4
·9451	37·05	44·10	77·3
·9452	37·00	44·05	77·2
·9453	36·95	44·00	77·1
·9454	36·90	43·95	77·0
·9455	36·85	43·90	76·9
·9456	36·75	43·85	76·8
·9457	36·70	43·80	76·7
·9458	36·65	43·70	76·6
·9459	36·60	43·65	76·5
·9460	36·55	43·60	76·4
·9461	36·50	43·55	76·3
·9462	36·45	43·50	76·2
·9463	36·40	43·40	76·1
·9464	36·35	43·35	76·0
·9465	36·30	43·30	75·9
·9466	36·25	43·25	75·8
·9467	36·20	43·20	75·7

I. Sp. gr. at 60° F. Water at 60°=1.	II. Percentage of alcohol *by weight*.	III. Percentage of alcohol *by volume*.	IV. Percentage of Proof Spirit.
·9468	36·15	43·15	75·6
·9469	36·10	43·10	75·5
·9470	36·05	43·00	75·4
·9471	36·00	42·95	75·3
·9472	35·90	42·90	75·2
·9473	35·85	42·85	75·1
·9474	35·80	42·80	75·0
·9475	35·75	42·75	74·9
·9476	35·70	42·65	74·7
·9477	35·65	42·60	74·6
·9478	35·60	42·55	74·5
·9479	35·55	42·45	74·4
·9480	35·50	42·40	74·3
·9481	35·45	42·30	74·2
·9482	35·40	42·25	74·1
·9483	35·35	42·20	74·0
·9484	35·30	42·15	73·9
·9485	35·25	42·10	73·8
·9486	35·20	42·05	73·7
·9487	35·15	42·00	73·6
·9488	35·10	41·95	73·5
·9489	35·05	41·90	73·4
·9490	35·00	41·85	73·3
·9491	34·95	41·75	73·2
·9492	34·85	41·70	73·0
·9493	34·80	41·65	72·9
·9494	34·75	41·55	72·8
·9495	34·70	41·50	72·7

I. Sp. gr. at 60° F. Water at 60°=1.	II. Percentage of alcohol *by weight*.	III. Percentage of alcohol *by volume*.	IV. Percentage of Proof Spirit.
·9496	34·65	41·40	72·6
·9497	34·55	41·35	72·5
·9498	34·50	41·30	72·4
·9499	34·45	41·25	72·3
·9500	34·40	41·20	72·2
·9501	34·35	41·15	72·1
·9502	34·30	41·10	72·0
·9503	34·25	41·00	71·8
·9504	34·20	40·95	71·7
·9505	34·15	40·85	71·6
·9506	34·10	40·80	71·5
·9507	34·00	40·75	71·4
·9508	33·95	40·70	71·2
·9509	33·90	40·65	71·1
·9510	33·85	40·55	71·0
·9511	33·80	40·50	70·9
·9512	33·75	40·40	70·8
·9513	33·70	40·35	70·7
·9514	33·65	40·30	70·6
·9515	33·60	40·25	70·5
·9516	33·50	40·20	70·4
·9517	33·45	40·10	70·3
·9518	33·40	40·05	70·2
·9519	33·35	40·00	70·1
·9520	33·30	39·90	70·0
·9521	33·25	39·85	69·8
·9522	33·20	39·80	69·7
·9523	33·10	39·70	69·6

I.	II.	III.	IV.
Sp. gr. at 60° F. Water at 60°=1.	Percentage of alcohol *by weight*.	Percentage of alcohol *by volume*.	Percentage of Proof Spirit.
·9524	33·05	39·65	69·5
·9525	33·00	39·55	69·4
·9526	32·95	39·50	69·3
·9527	32·90	39·40	69·2
·9528	32·80	39·35	69·0
·9529	32·75	39·30	68·9
·9530	32·70	39·20	68·7
·9531	32·65	39·15	68·6
·9532	32·60	39·10	68·5
·9533	32·50	39·05	68·4
·9534	32·45	38·95	68·3
·9535	32·40	38·90	68·2
·9536	32·35	38·85	68·1
·9537	32·30	38·80	68·0
·9538	32·25	38·75	67·9
·9539	32·20	38·70	67·8
·9540	32·15	38·60	67·6
·9541	32·10	38·55	67·5
·9542	32·00	38·45	67·4
·9543	31·95	38·40	67·3
·9544	31·90	38·35	67·2
·9545	31·85	38·30	67·1
·9546	31·80	38·20	67·0
·9547	31·70	38·15	66·8
·9548	31·65	38·05	66·7
·9549	31·60	38·00	66·6
·9550	31·55	37·95	66·5
·9551	31·50	37·90	66·4

I. Sp. gr. at 60° F. Water at 60°=1.	II. Percentage of alcohol *by weight*.	III. Percentage of alcohol *by volume*.	IV. Percentage of Proof Spirit.
·9552	31·45	37·85	66·3
·9553	31·40	37·80	66·2
·9554	31·35	37·75	66·1
·9555	31·30	37·70	66·0
·9556	31·25	37·65	65·9
·9557	31·20	37·55	65·8
·9558	31·10	37·50	65·6
·9559	31·05	37·40	65·5
·9560	31·00	37·35	65·4
·9561	30·90	37·25	65·3
·9562	30·85	37·15	65·1
·9563	30·80	37·10	65·0
·9564	30·70	37·00	64·9
·9565	30·65	36·95	64·7
·9566	30·60	36·85	64·6
·9567	30·50	36·75	64·4
·9568	30·45	36·70	64·3
·9569	30·40	36·65	64·2
·9570	30·35	36·55	64·1
·9571	30·30	36·50	64·0
·9572	30·25	36·45	63·8
·9573	30·20	36·40	63·7
·9574	30·10	36·30	63·6
·9575	30·05	36·20	63·5
·9576	30·00	36·15	63·4
·9577	29·90	36·05	63·2
·9578	29·85	36·00	63·1
·9579	29·80	35·90	62·9
·9580	29·70	35·85	62·8

I. Sp. gr. at 60° F. Water at 60°=1.	II. Percentage of alcohol *by weight*.	III. Percentage of alcohol *by volume*.	IV. Percentage of Proof Spirit.
·9581	29·65	35·80	62·7
·9582	29·60	35·70	62·5
·9583	29·55	35·60	62·4
·9584	29·50	35·55	62·3
·9585	29·40	35·45	62·2
·9586	29·35	35·40	62·0
·9587	29·25	35·30	61·9
·9588	29·20	35·25	61·8
·9589	29·15	35·20	61·7
·9590	29·10	35·10	61·6
·9591	29·00	35·05	61·5
·9592	28·95	35·00	61·3
·9593	28·90	34·90	61·2
·9594	28·85	34·80	61·1
·9595	28·75	34·75	60·9
·9596	28·65	34·65	60·8
·9597	28·60	34·60	60·7
·9598	28·55	34·55	60·5
·9599	28·50	34·45	60·4
·9600	28·45	34·40	60·3
·9601	28·40	34·40	60·1
·9602	28·30	34·30	59·9
·9603	28·25	34·20	59·8
·9604	28·20	34·10	59·7
·9605	28·10	34·00	59·6
·9606	28·05	33·95	59·5
·9607	28·00	33·85	59·4
·9608	27·90	33·75	59·3
·9609	27·85	33·70	59·1

I. Sp. gr. at 60° F. Water at 60°=1.	II. Percentage of alcohol *by weight*.	III. Percentage of alcohol *by volume*.	IV. Percentage of Proof Spirit.
·9610	27·80	33·60	59·0
·9611	27·75	33·55	58·8
·9612	27·65	33·45	58·7
·9613	27·60	33·40	58·5
·9614	27·55	33·35	58·4
·9615	27·45	33·25	58·3
·9616	27·40	33·20	58·1
·9617	27·30	33·10	58·0
·9618	27·25	33·05	57·9
·9619	27·20	33·00	57·8
·9620	27·15	32·90	57·6
·9621	27·10	32·80	57·5
·9622	27·05	32·70	57·4
·9623	27·00	32·65	57·2
·9624	26·90	32·55	57·1
·9625	26·80	32·45	56·9
·9626	26·75	32·40	56·8
·9627	26·65	32·30	56·6
·9628	26·60	32·20	56·5
·9629	26·50	32·15	56·3
·9630	26·45	32·05	56·2
·9631	26·35	31·95	56·0
·9632	26·30	31·90	55·9
·9633	26·25	31·80	55·8
·9634	26·15	31·75	55·6
·9635	26·10	31·65	55·5
·9636	26·00	31·55	55·3
·9637	25·95	31·50	55·2

I. Sp. gr. at 60° F. Water at 60°=1.	II. Percentage of alcohol *by weight.*	III. Percentage of alcohol *by volume.*	IV. Percentage of Proof Spirit.
·9638	25·85	31·40	55·0
·9639	25·75	31·30	54·8
·9640	25·70	31·20	54·7
·9641	25·65	31·15	54·5
·9642	25·60	31·05	54·4
·9643	25·50	31·00	54·2
·9644	25·45	30·90	54·1
·9645	25·40	30·85	53·9
·9646	25·30	30·75	53·8
·9647	25·25	30·70	53·6
·9648	25·15	30·60	53·5
·9649	25·10	30·50	53·4
·9650	25·00	30·40	53·3
·9651	24·90	30·30	53·1
·9652	24·85	30·20	53·0
·9653	24·75	30·10	52·8
·9654	24·70	30·00	52·6
·9655	24·60	29·95	52·5
·9656	24·55	29·85	52·3
·9657	24·50	29·80	52·2
·9658	24·40	29·70	52·0
·9659	24·35	29·65	51·9
·9660	24·25	29·55	51·8
·9661	24·20	29·45	51·6
·9662	24·15	29·40	51·5
·9663	24·05	29·30	51·3
·9664	24·00	29·20	51·2
·9665	23·90	29·10	51·0

I. Sp. gr. at 60° F. Water at 60°=1.	II. Percentage of alcohol *by weight*.	III. Percentage of alcohol *by volume*.	IV. Percentage of Proof Spirit.
·9666	23·80	29·00	50·8
·9667	23·75	28·90	50·7
·9668	23·65	28·80	50·5
·9669	23·55	28·70	50·3
·9670	23·50	28·65	50·2
·9671	23·40	28·55	50·0
·9672	23·35	28·45	49·9
·9673	23·25	28·35	49·7
·9674	23·20	28·25	49·5
·9675	23·10	28·15	49·3
·9676	23·05	28·10	49·2
·9677	23·00	28·05	49·1
·9678	22·90	27·95	49·0
·9679	22·85	27·85	48·8
·9680	22·75	27·75	48·6
·9681	22·65	27·65	48·5
·9682	22·60	27·55	48·3
·9683	22·50	27·50	48·2
·9684	22·45	27·40	48·0
·9685	22·35	27·30	47·8
·9686	22·25	27·20	47·6
·9687	22·20	27·10	47·4
·9688	22·10	27·00	47·2
·9689	22·00	26·90	47·0
·9690	21·95	26·85	46·9
·9691	21·85	26·75	46·8
·9692	21·80	26·65	46·6
·9693	21·75	26·55	46·5

I. Sp. gr. at 60° F. Water at 60°=1.	II. Percentage of alcohol *by* *weight*.	III. Percentage of alcohol *by* *volume*.	IV. Percentage of Proof Spirit.
·9694	21·65	26·45	46·3
·9695	21·55	26·35	46·2
·9696	21·45	26·25	46·0
·9697	21·35	26·10	45·8
·9698	21·30	26·00	45·6
·9699	21·20	25·90	45·4
·9700	21·10	25·75	45·2
·9701	21·00	25·65	45·0
·9702	20·90	25·60	44·9
·9703	20·85	25·50	44·7
·9704	20·80	25·40	44·6
·9705	20·70	25·35	44·4
·9706	20·65	25·25	44·3
·9707	20·55	25·15	44·1
·9708	20·50	25·05	44·0
·9709	20·45	25·00	43·8
·9710	20·35	24·90	43·6
·9711	20·25	24·80	43·4
·9712	20·20	24·70	43·2
·9713	20·10	24·60	43·0
·9714	20·00	24·50	42·9
·9715	19·90	24·40	42·7
·9716	19·85	24·30	42·5
·9717	19·75	24·20	42·4
·9718	19·70	24·10	42·3
·9719	19·60	24·00	42·1
·9720	19·55	23·90	42·0
·9721	19·45	23·85	41·8

I. Sp. gr. at 60° F. Water at 60°=1.	II. Percentage of alcohol *by weight*.	III. Percentage of alcohol *by volume*.	IV. Percentage of Proof Spirit.
·9722	19·40	23·80	41·7
·9723	19·30	23·70	41·6
·9724	19·25	23·65	41·5
·9725	19·20	23·60	41·3
·9726	19·15	23·50	41·2
·9727	19·10	23·45	41·1
·9728	19·00	23·30	40·9
·9729	18·90	23·20	40·7
·9730	18·80	23·05	40·5
·9731	18·70	22·95	40·3
·9732	18·60	22·80	40·1
·9733	18·50	22·70	39·9
·9734	18·40	22·55	39·7
·9735	18·30	22·45	39·5
·9736	18·20	22·35	39·3
·9737	18·15	22·25	39·1
·9738	18·05	22·15	38·9
·9739	17·95	22·05	38·7
·9740	17·90	21·95	38·5
·9741	17·80	21·85	38·3
·9742	17·75	21·80	38·1
·9743	17·65	21·70	37·9
·9744	17·55	21·60	37·8
·9745	17·50	21·50	37·6
·9746	17·40	21·40	37·4
·9747	17·35	21·30	37·3
·9748	17·25	21·20	37·1
·9749	17·20	21·10	37·0
·9750	17·10	21·00	36·8

I. Sp. gr. at 60° F. Water at 60°=1.	II. Percentage of alcohol *by* *weight*.	III. Percentage of alcohol *by* *volume*.	IV. Percentage of Proof Spirit.
·9751	17·00	20·90	36·6
·9752	16·95	20·80	36·4
·9753	16·85	20·70	36·3
·9754	16·75	20·60	36·1
·9755	16·65	20·50	36·0
·9756	16·60	20·40	35·8
·9757	16·55	20·30	35·6
·9758	16·45	20·20	35·4
·9759	16·40	20·10	35·2
·9760	16·30	20·00	35·1
·9761	16·20	19·90	34·9
·9762	16·10	19·80	34·7
·9763	16·05	19·70	34·6
·9764	15·95	19·60	34·4
·9765	15·90	19·50	34·2
·9766	15·80	19·40	34·0
·9767	15·70	19·30	33·8
·9768	15·60	19·20	· 33·7
·9769	15·50	19·10	*v* 33·5
·9770	15·45	19·00	33·3
·9771	15·40	18·90	32·2
·9772	15·30	18·80	33·0
·9773	15·20	18·70	32·8
·9774	15·10	18·60	32·6
·9775	15·00	18·50	32·4
·9776	14·95	18·40	32·2
·9777	14·85	18·30	32·0
·9778	14·80	18·20	31·9
·9779	14·70	18·10	31·7

F

I. Sp. gr. at 60° F. Water at 60°=1.	II. Percentage of alcohol *by weight*.	III. Percentage of alcohol *by volume*.	IV. Percentage of Proof Spirit.
·9780	14·65	18·00	31·6
·9781	14·60	17·90	31·4
·9782	14·50	17·80	31·2
·9783	14·40	17·70	31·1
·9784	14·30	17·60	30·9
·9785	14·25	17·55	30·8
·9786	14·15	17·45	30·6
·9787	14·05	17·30	30·4
·9788	13·95	17·20	30·2
·9789	13·90	17·10	30·0
·9790	13·80	17·00	29·8
·9791	13·70	16·90	29·6
·9792	13·60	16·80	29·4
·9793	13·55	16·70	29·3
·9794	13·45	16·60	29·1
·9795	13·40	16·50	28·9
·9796	13·30	16·40	28·7
·9797	13·25	16·30	28·6
·9798	13·15	16·20	28·4
·9799	13·10	16·10	28·2
·9800	13·00	16·00	28·0
·9801	12·90	15·90	27·8
·9802	12·85	15·80	27·7
·9803	12·75	15·70	27·6
·9804	12·65	15·65	27·4
·9805	12·60	15·55	27·3
·9806	12·55	15·45	27·1
·9807	12·50	15·35	27·0

I. Sp. gr. at 60° F. Water at 60°=1.	II. Percentage of alcohol *by weight*.	III. Percentage of alcohol *by volume*.	IV. Percentage of Proof Spirit.
·9808	12·40	15·25	26·8
·9809	12·30	15·20	26·6
·9810	12·25	15·10	26·5
·9811	12·15	15·00	26·3
·9812	12·10	14·95	26·2
·9813	12·00	14·85	26·0
·9814	11·90	14·75	25·8
·9815	11·85	14·65	25·6
·9816	11·75	14·55	25·4
·9817	11·70	14·45	25·3
·9818	11·60	14·35	25·1
·9819	11·50	14·25	24·9
·9820	11·45	14·10	24·7
·9821	11·35	14·00	24·5
·9822	11·25	13·90	24·4
·9823	11·20	13·80	24·2
·9824	11·10	13·70	24·0
·9825	11·05	13·65	23·9
·9826	10·95	13·55	23·8
·9827	10·90	13·45	23·6
·9828	10·80	13·35	23·4
·9829	10·75	13·25	23·3
·9830	10·65	13·20	23·1
·9831	10·60	13·10	23·0
·9832	10·50	13·00	22·8
·9833	10·40	12·90	22·7
·9834	10·35	12·85	22·5
·9835	10·30	12·75	22·4

I. Sp. gr. at 60° F. Water at 60°=1.	II. Percentage of alcohol *by weight*.	III. Percentage of alcohol *by volume*.	IV. Percentage of Proof Spirit.
·9836	10·20	12·65	22·2
·9837	10·15	12·60 .	22·0
·9838	10·05	12·50	21·9
·9839	10·00	12·40	21·7
·9840	9·90	12·35	21·5
·9841	9·85	12·25	21·3
·9842	9·80	12·15	21·2
·9843	9·75	12·05	21·0
·9844	9·70	11·95	20·9
·9845	9·55	11·85	20·7
·9846	9·50	11·80	20·6
·9847	9·40	11·70	20·5
·9848	9·35	11·60	20·3
·9849	9·25	11·50	20·2
·9850	9·20	11·40	20·0
·9851	9·10	11·30	19·8
·9852	9·05	11·20	19·7
·9853	9·00	11·10	19·5
·9854	8·90	11·00	19·3
·9855	8·85	10·95	19·2
·9856	8·75	10·85	19·0
·9857	8·70	10·80	18·9
·9858	8·60	10·75	18·7
·9859	8·55	10·65	18·5
·9860	8·50	10·55	18·4
·9861	8·40	10·45	18·2
·9862	8·35	10·35	18·0
·9863	8·30	10·25	17·9

I.	II.	III.	IV.
Sp. gr. at 60° F. Water at 60°=1.	Percentage of alcohol *by weight*.	Percentage of alcohol *by volume*.	Percentage of Proof Spirit.
·9864	8·20	10·15	17·7
·9865	8·10	10·05	17·6
·9866	8·05	10·00	17·5
·9867	8·00	9·90	17·4
·9868	7·90	9·85	17·2
·9869	7·85	9·75	17·1
·9870	7·80	9·65	16·9
·9871	7·70	9·60	16·8
·9872	7·65	9·50	16·6
·9873	7·55	9·45	16·5
·9874	7·50	9·35	16·4
·9875	7·45	9·25	16·2
·9876	7·35	9·15	16·0
·9877	7·30	9·10	15·9
·9878	7·25	9·00	15·8
·9879	7·15	8·90	15·7
·9880	7·10	8·80	15·5
·9881	7·00	8·75	15·4
·9882	6·95	8·65	15·2
·9883	6·90	8·60	15·1
·9884	6·80	8·50	14·9
·9885	6·70	8·40	14·8
·9886	6·65	8·35	14·7
·9887	6·60	8·30	14·5
·9888	6·55	8·15	14·4
·9889	6·50	8·10	14·2
·9890	6·40	8·00	14·1
·9891	6·35	7·95	13·9

I. Sp. gr. at 60° F. Water at 60° = 1.	II. Percentage of alcohol *by weight*.	III. Percentage of alcohol *by volume*.	IV. Percentage of Proof Spirit.
·9892	6·30	7·85	13·8
·9893	6·20	7·80	13·7
·9894	6·15	7·70	13·5
·9895	6·10	7·60	13·4
·9896	6·00	7·55	13·2
·9897	5·95	7·45	13·1
·9898	5·90	7·40	12·9
·9899	5·85	7·30	12·8
·9900	5·75	7·15	12·6
·9901	5·70	7·05	12·5
·9902	5·65	7·00	12·3
·9903	5·60	6·95	12·2
·9904	5·50	6·85	12·0
·9905	5·45	6·80	11·9
·9906	5·40	6·75	11·8
·9907	5·30	6·70	11·6
·9908	5·25	6·60	11·5
·9909	5·20	6·50	11·4
·9910	5·15	6·40	11·2
·9911	5·05	6·30	11·0
·9912	5·00	6·20	10·9
·9913	4·95	6·15	10·8
·9914	4·90	6·10	10·7
·9915	4·80	6·00	10·5
·9916	4·75	5·95	10·4
·9917	4·70	5·90	10·3
·9918	4·65	5·80	10·2
·9919	4·55	5·70	10·0
·9920	4·50	5·65	9·9

I. Sp. gr. at 60° F. Water at 60°=1.	II. Percentage of alcohol *by weight*.	III. Percentage of alcohol *by volume*.	IV. Percentage of Proof Spirit.
·9921	4·45	5·55	9·8
·9922	4·40	5·50	9·6
·9923	4·35	5·40	9·5
·9924	4·25	5·30	9·4
·9925	4·20	5·25	9·2
·9926	4·15	5·20	9·1
·9927	4·10	5·15	9·0
·9928	4·00	5·05	8·9
·9929	3·95	5·00	8·8
·9930	3·90	4·90	8·6
·9931	3·85	4·85	8·5
·9932	3·80	4·80	8·3
·9933	3·75	4·70	8·2
·9934	3·65	4·65	8·1
·9935	3·60	4·55	7·9
·9936	3·55	4·50	7·8
·9937	3·50	4·45	7·6
·9938	3·40	4·30	7·5
·9939	3·35	4·25	7·3
·9940	3·30	4·15	7·2
·9941	3·25	4·10	7·1
·9942	3·20	4·00	7·0
·9943	3·15	3·95	6·9
·9944	3·10	3·85	6·7
·9945	3·00	3·80	6·6
·9946	2·95	3·75	6·5
·9947	2·90	3·70	6·4
·9948	2·85	3·60	6·3
·9949	2·80	3·55	6·2

I. Sp. gr. at 60° F. Water at 60°=1.	II. Percentage of alcohol *by weight*.	III. Percentage of alcohol *by volume*.	IV. Percentage of Proof Spirit.
˙9950	2˙75	3˙50	6˙1
˙9951	2˙70	3˙40	5˙9
˙9952	2˙60	3˙30	5˙7
˙9953	2˙55	3˙20	5˙6
˙9954	2˙50	3˙15	5˙5
˙9955	2˙45	3˙10	5˙4
˙9956	2˙40	3˙00	5˙3
˙9957	2˙35	2˙90	5˙2
˙9958	2˙30	2˙85	5˙1
˙9959	2˙20	2˙80	5˙0
˙9960	2˙15	2˙70	4˙9
˙9961	2˙10	2˙65	4˙7
˙9962	2˙05	2˙60	4˙6
˙9963	2˙00	2˙50	4˙4
˙9964	1˙95	2˙45	4˙3
˙9965	1˙90	2˙40	4˙1
˙9966	1˙85	2˙30	4˙0
˙9967	1˙80	2˙20	3˙9
˙9968	1˙75	2˙15	3˙8
˙9969	1˙65	2˙05	3˙6
˙9970	1˙60	2˙00	3˙5
˙9971	1˙55	1˙95	3˙4
˙9972	1˙50	1˙85	3˙3
˙9973	1˙45	1˙80	3˙2
˙9974	1˙40	1˙75	3˙1
˙9975	1˙35	1˙70	3˙0
˙9976	1˙30	1˙65	2˙8
˙9977	1˙25	1˙55	2˙7

I. Sp. gr. at 60° F. Water at 60°=1.	II. Percentage of alcohol *by weight.*	III. Percentage of alcohol *by volume.*	IV. Percentage of Proof Spirit.
·9978	1·20	1·45	2·6
·9979	1·10	1·40	2·5
·9980	1·05	1·30	2·4
·9981	1·00	1·25	2·2
·9982	0·95	1·20	2·0
·9983	0·90	1·10	1·9
·9984	0·85	1·05	1·8
·9985	0·80	1·00	1·7
·9986	0·75	0·90	1·6
·9987	0·70	0·85	1·5
·9988	0·65	0·80	1·4
·9989	0·60	0·70	1·3
·9990	0·55	0·65	1·2
·9991	0·45	0·55	1·0
·9992	0·40	0·50	0·9
·9993	0·35	0·45	0·8
·9994	0·30	0·40	0·7
·9995	0·25	0·30	0·6
·9996	0·20	0·25	0·5
·9997	0·15	0·20	0·4
·9998	0·10	0·15	0·2
·9999	0·05	0·05	0·1

73